# DARK TOURISM AND RURAL CRIME

# Research in Rural Crime series

*Series editors:* **Alistair Harkness**, University of New England in New South Wales, Australia and **Jessica René Peterson**, Southern Oregon University, US

---

The *Research in Rural Crime* series reflects the multi-faceted nature of rural crime and provides an outlet for original, cutting-edge research in this emergent criminological subfield. Truly international in nature, it leads the way for new thinking on a wide range of rural crime topics, rural transgressions, security and justice.

## Find out more about the new and forthcoming titles in the series:

bristoluniversitypress.co.uk/research-in-rural-crime

# DARK TOURISM AND RURAL CRIME

## Crime and Punishment in Rural Australia

Jenny Wise

BRISTOL
UNIVERSITY
PRESS

First published in Great Britain in 2024 by

Bristol University Press
University of Bristol
1–9 Old Park Hill
Bristol
BS2 8BB
UK
t: +44 (0)117 374 6645
e: bup-info@bristol.ac.uk

Details of international sales and distribution partners are available at bristoluniversitypress.co.uk

© Bristol University Press 2024

British Library Cataloguing in Publication Data
A catalogue record for this book is available from the British Library

ISBN 978-1-5292-1925-8 hardcover
ISBN 978-1-5292-1926-5 ePub
ISBN 978-1-5292-1927-2 ePdf

Cover design: Liam Roberts Design
Front cover image: alamy/mtphoto19
Bristol University Press uses environmentally responsible print partners.
Printed and bound in Great Britain by CPI Group (UK) Ltd, Croydon, CR0 4YY

FSC
www.fsc.org
MIX
Paper | Supporting
responsible forestry
FSC® C013604

For Nathan, Tilly and Lachlan. Thank you for always inspiring, supporting and being patient with my dark tourism adventures.

# Contents

# Series Editors' Preface

*Alistair Harkness and Jessica René Peterson*

Contemporary criminology grew out of mass transformations in society during the twentieth century – a period that witnessed the formation and consolidation of cities through migration and the restructuring of 'the urban' following the emergence of the information age. Considerable debate, research and scholarly theory have been formed primarily within the urban domain in this regard: witness, for example, the influence of the Chicago School and its influence on the rise of urban ecology-based approaches to crime prevention.

Rural areas themselves are currently being transformed by the new patterns of global flows as societies undergo transitions within. Nevertheless, myths about peaceful, crime-free areas beyond the cityscape persist, when, in fact, rural crime is multi-faceted – raising new policy predicaments about policing and security governance. With approximately 45 per cent of the global population living in rural areas, a focus on rural crime in these diverse communities is critical. The series provides a space for new research and writing on a wide range of rural crime topics, rural transgressions, security and justice.

The *Research in Rural Crime* series has emerged to fill an important gap – to provide an outlet for mid-length monographs which focus on rural crime and responses to rural crime – providing an opportunity for original, cutting-edge monograph-length research in the criminological subfield of rural criminology. Truly international, it welcomes and produces titles that are jurisdictional specific or related to themes that transcend political and juridical boundaries, and presents outlooks on contemporary theory, research and pressing public policy issues.

In essence, this series provides a formalised space for voices hitherto overlooked or without a venue for longer-length exploration of rural crime, policing, security and other issues. It allows for the consolidation of intellectual thoughtfulness in monograph form, either from sole or joint-authored volumes or from groups of colleagues in edited collections. Importantly, too, it provides an opportunity for the combination of academic scholarship and empirical research with contemporary application.

Contributors to this series provide cutting-edge interdisciplinary and comparative rural criminological perspectives. Titles will be theoretically and conceptually driven, empirical or adopting mixed-methods approaches, and topics will focus on regional, rural and remote parts of the globe that are often overlooked in criminological works.

## Editors' Foreword to *Dark Tourism and Rural Crime*

This volume from Jenny Wise offers a gripping and intriguing assessment of the travel to sites that represent death, disaster or atrocity known as 'dark tourism'. With a focus on Australia, Wise also incorporates international examples to illustrate the key dimensions of this popular socio-cultural phenomenon. Drawing upon a strong secondary evidence base, the book benefits enormously from original field work.

There can be no doubt as to the allure of places of horror and trauma for many, and thus the economic motivations for the formalisation of dark tourism. But even where there is no fixed place, such as in the Belanglo Forest in New South Wales, or formalised tourism industry, such as in Wittenoom in Western Australia, a morbid fascination with the grisly, macabre and outright criminal (whether perpetrated by individuals, industry or government) will continue to attract attention. Consequently, there is a much-needed scholarly focus, and this book delivers precisely that.

Australian readers and those familiar with the country will immediately recognise many of the examples which are detailed throughout the book. Indeed, and for example, prior to the township of Glenrowan being bypassed in the late 1980s, anyone driving along the Hume Highway between Melbourne and Sydney would have encountered first-hand the industry which emerged around the infamous Ned Kelly. Even in the third decade of the twenty-first century, this place and the mythology of this anti-establishment folk hero – or 'cop killer' depending on one's perspective – is unavoidable in the national psyche, and encourages a detour from the Highway.

However, readers from other locations should not fear any unfamiliarity, because Wise adeptly contextualises key examples to solidify points made about the development and maintenance of dark tourism destinations internationally. She explores the historical and cultural perspectives, and contemporary controversies surrounding, for example, the juxtaposing of memorialising versus sensationalising crimes in rural places.

The book charts a course from early colonial violence and tourism in rural places to more contemporary examples, and concludes with reflections on the significance and future of rural dark tourism. This book fills an empty gap by enhancing our knowledge of the various dimensions of dark tourism and will serve as an exemplar for scholars wishing to consider similar rural-focused studies in other places internationally. Academics, students and the macabre- or oddities-interested reader alike will take away much from this excellent volume.

# List of Figures

# List of Abbreviations

| | |
|---|---|
| ACRE | Australian Centre for Rural Entrepreneurship |
| APPI | Australian Paranormal Phenomenon Investigators |
| NSW | New South Wales |
| PAHS | Port Arthur Historic Site |
| PAHSMA | Port Arthur Historic Site Management Authority |
| SPB | Scenery Preservation Board |

# About the Author

Jenny Wise is Associate Professor in Criminology in the School of Humanities, Arts and Social Sciences at the University of New England, located in New South Wales, Australia. Her research focuses on dark tourism, crime as a form of leisure and the social impacts of forensic science on the criminal justice system. However, a love of pop culture sometimes diverts her research into new and interesting spaces.

# Acknowledgements

As with any endeavour, this book would not have been possible without the help and support of numerous people. Thank you to the series editors, Alistair Harkness and Jessica René Peterson (and a previous series editor, Matt Bowden), for taking an interest in my research, and to Rebecca Tomlinson, Grace Carroll, Angela Gage, Freya Trand, Sarah Green and Emma Cook from Bristol University Press for all your help throughout this process.

I offer a heartfelt gratitude to an amazing friend, and my frequent dark tourism co-author, Lesley McLean, who has inspired and helped me in all my endeavours. Similarly, I am very grateful for the friendship and constant support offered by Kyle Mulrooney, and of course, his push to make me write the book proposal and submit it. Thank you to my parents, Anne and Neil, for always believing and loving me – sorry I missed seeing and calling you more frequently while writing this book!

Most importantly, I couldn't have accomplished any of this work without the love and support of my husband, Nathan, and our two wonderful children, Tilly and Lachlan. The three of you constantly inspire me, and your understanding and willingness to be dragged around the countryside in pursuit of my research is appreciated more than you know.

# Preface

There is an enduring public interest in visiting sites of death, tragedy and suffering as part of a formalised tour or experience (Casella and Fennelly, 2016). Attached to creating and visiting these sites is a myriad of issues intertwined: the purpose of the site, motivations of tourists, interpretation and community acceptance or resistance. As White (2016, para. 2) aptly states, 'the relationship between trauma, tourism, commemoration and the nature of the place itself is a complicated one'. Because of these complexities, it is a very important area of study, and in particular, a comprehensive examination of dark tourism in rural areas in Australia is well overdue.

At a time when rural and regional communities are turning specifically towards dark tourist activities to keep their community economically viable, it is essential that this phenomenon is examined and the distinctive constraints and opportunities for rural communities are understood. It is also necessary to understand the role that dark tourism plays in creating cultural narratives of Australia's rural and regional spaces.

As such, this book is divided into two main sections. Part I explores rural dark tourism associated with Australia's colonial history, including colonial violence, convict tourism and bushrangers. Part II presents an analysis of more 'modern' crime tourist sites and includes carceral tourism, serial killers and sensational crimes and deadly towns, ghosts, crime and bushfires. In essence, the book provides an exploration of the range of dark tourism activities available in rural and regional Australia around crime and punishment using a rural lens of analysis.

*Jenny Wise*
*March 2024*

# Introduction

One of Australia's earliest cases of 'dark tourism' involved spectators travelling from as far as Melbourne to witness the bushrangers known as the 'Kelly Gang' during their 'last stand' against police officers at Glenrowan in rural Victoria. On 27 June 1880 Ned Kelly, Dan Kelly, Joe Byrne and Steve Hart held 60 hostages in the Glenrowan Hotel in the hopes of ambushing police officers. At 3 am the following morning, police surrounded the hotel and gunfire ensued between the Kelly Gang and the police, killing some of the hostages in the building. The hotel was burnt to the ground, causing further fatalities.

The police, anticipating the siege, invited four Victorian journalists to document what they hoped would be the capture of Ned Kelly and his criminal associates. The inclusion of the reporters meant that the siege, accompanied by photographs, became international news. The news coverage also illustrated the desire of people to witness the macabre, including 'real-life' crime unfolding. For example, the *Town and Country Journal* (1880a, 8) reported that there were 'thousands of people' at Glenrowan from different parts of the country. According to the media, 'great excitement prevail[ed]' at Glenrowan with 'numbers of people constantly arriving' on horseback to witness the siege (Town and Country Journal, 1880b, 8). Other spectators arrived on the midday train from as far as Melbourne to see the 'entertainment' or 'piece of theatre' (Cochrane, 2022, para. 10). Indeed, Cochrane (2022, para. 10), writing to stimulate school history students to analyse historical images, argues that the spectators behaved like an audience attending the theatre, 'cheer[ing] and clapp[ing] at the most dramatic moments'.

Following the siege, those who had travelled searched for 'keepsakes' or 'souvenirs' of the experience:

> The ruins of the hotel were scanned eagerly, and all the relics in the shape of knives, forks, bullets, empty cartridge cases, were seized upon. Some of the bullets lodged in the stockyard fence, and these were at

once cut out and appropriated. The spot where Ned Kelly fell was also a subject of great interest, and some leaves with blood upon them were taken away as treasures. (Town and Country Journal, 1880a, 8)

The photographs taken during the siege, and of the wounded and dead bushrangers after the siege, also became highly sought-after mementos. The rarity of available photos led to the creation of 'fake photographs' and sketches, while other tourists decided to take their own pictures in iconic siege locations. For example, one historical photograph shows a group of five men and a child on the 'Ned Kelly log' (the place where he collapsed and was apprehended by police) taken two days after Kelly himself lay there (Chen, 2018, 11). A year later on 9 September 1881, the media reported that 'a large gathering of Benalla state school children' travelled to Glenrowan to picnic at the 'battle-field' of the Kelly gang and the police (Australasian Sketcher, 1881, 305). The article concluded that the picnic of 3,000 students and teachers was a great success and that 'the children seemed quite alive to the historical associations of the place, and all spent the day very agreeably' (Australasian Sketcher, 1881, 305). As will be explored in Chapter 4 of this book, tourism to Ned Kelly country continues to this day, with a new array of souvenirs available for purchase as well as new photo opportunities.

Dark tourism continues in modern times, perhaps becoming more prominent with the increased ease of travel and the swiftness with which people find out about 'dark' incidents and locations. For example, in May 1999, the small rural South Australian town of Snowtown made headlines locally, nationally and internationally after the police located the remains of eight bodies stored in six 44-gallon barrels filled with acid inside a local disused bank. As will be detailed in Chapter 6, journalistic embellishments led to the conversion of an idyllic town to a dystopian destination for dark tourists. Media accounts advised the national community that 'ghoulish tourists' began to arrive the weekend after the discovery to take photos and 'sniff' around the bank. The media's description of tourists to Snowtown reinforces notions that travelling to sites of death and pain is indeed 'dark' and that only certain people would be attracted to such sites.

What, then, is 'dark tourism'? Dark tourism involves travel to sites that provide representations of death, disaster or atrocity. In recent years, dark tourism has been growing in rural areas as communities have actively adopted a 'dark tourism strategy' to encourage visitors to the area. Yet, in other areas, communities have sought to distance themselves from 'rural dystopias' and what is perceived to be unethical tourism practices. Within Australia, there is a range of rural dark tourism activities available for tourists including convict and penal museums, 'deadly towns', sites of serial killings, sites of colonial violence, locations associated with bushrangers, crime and ghost tours and even the potential for tourism of natural disaster sites. This book uses specific case studies of these different types of dark tourism sites to explore the unique considerations and constraints of tourism within rural and regional Australia, and how such sites contribute to Australia's cultural heritage and national identity with a particular focus on concepts of the rural idyll and rural dystopias.

Dark tourism continues to grow in popularity, with Sampson noting that 'millions of tourists around the world ... [visit] some of the most unhappiest places on Earth' (Sampson, 2019, para. 1). Coupled with this is increased public recognition of and reflection on the culture, which has been aided by shows such as *Dark Tourist* on the streaming platform Netflix.

Around the world, interdisciplinary scholars have sought to understand this growing phenomenon and its impact on society. Their research has many other names, including 'negative sightseeing' (MacCannell, 1989), 'Black Spots tourism' (Rojek, 1993), 'thanatourism' (Seaton, 1996), 'tragic tourism' (Lippard, 1999), 'grief tourism' (O'Neill, 2002) and 'fright tourism' (Bristow and Newman, 2004).

But, while dark tourism is gaining momentum as a field of study, there are notable gaps in the literature. One gap is that most academic studies (and media attention) focus on notorious sites of death, disaster and atrocity, such as Holocaust museums and sites, famous decommissioned prisons such as Alcatraz in the United States, Robben Island in South Africa and The Clink in Britain (Piché and Walby, 2018), or sites where famous criminals such as Myra Hindley in the United Kingdom and Charles Manson in the United States operated. Smaller, less well-known sites are often overlooked in these studies, and yet those smaller sites provide experiences that are just as rich as many of their more well-known counterparts. Further, courthouses, police museums and lockups are often overlooked in scholarly research, yet they remain important as they continue to shape and influence public memory and popular history. However, the largest gap with existing studies is the absence of exploring the importance of geographical variances among dark tourist sites, in particular, the difference between metropolitan and rural/ regional sites.

## Dark tourism and cultural criminology

The term 'dark tourism' was coined by Lennon and Foley (2000) and broadly refers to travel to tourist sites that provide 'representations of death, disaster or atrocity for pedagogical and commercial purposes' (Walby and Piché, 2011, 452). The fascination that humans have 'with our ability to do evil, witness the evidence of horror and stare fixedly at photographic, filmic or heritage artefacts connected with death is at the heart of these phenomena known as "Dark Tourism"' (Lennon, 2017, 217).

For Lennon and Foley (2000, 11) there are three critical features of dark tourism: first, global communication is essential in creating interest in the site; second, the objects at the site need to create anxiety and doubt over modernity; and third, 'the educative elements of sites are accompanied by elements of commodification and a commercial ethic which (whether explicit or implicit) accepts that visitation (whether purposive or incidental) is an opportunity to develop a tourism product'. A number of problems and limitations have been identified with these categorisations. For example, under this classification, sites of death that occurred prior to the start of the twentieth century cannot be classified as dark tourist destinations. According to Lennon and Foley (2000, 12), such places are sites of tourism, but because they lack the 'anxiety and doubt within interpretation offered and the design of the sites as both products and experiences (including merchandising and revenue generation) that introduced "dark tourism"' they are not 'dark' tourist sites. Additionally, because such atrocities occurred outside of living memory such sites do not cause anxiety and doubt about modernity and its consequences, which Lennon and Foley (2000) believe to be essential for a site to be a dark tourist destination.

Since 2000, there has been academic debate over this categorisation and what actually constitutes a 'dark tourism' site. Many academics argue that the tourist niche for visiting places of tragedy, death, destruction and atrocity has a long historical past and that this still constitutes 'dark tourism' activity. For example, Stone (2006) provides several historical examples of dark tourism. In 1838 there was an alleged guided railway tour in Cornwall, England to view the hanging of two convicted murderers. Guided morgue tours were conducted in the Victorian period (see Figure 1.1), and throughout the nineteenth century, there were galleries built in 'correction houses' to 'accommodate fee-paying visitors who witnessed flogging as a recreational activity' (Stone, 2006, 147).

While there is debate over historical sites being labelled as 'dark tourism' sites, there is widespread agreement that 'dark tourism over the last century has become more widespread and varied' (Stone, 2006, 147). Further, as a society, we seem to be more concerned about 'dark tourism' today than we have been in the past and how the 'postmodern condition' influences travel. In a way, travel to sites of death and atrocity has been 'revived', 'retrieved' and 'rediscovered' with heritage tourism now becoming trendy, globalised, mediatised and sold as a global commodity (Stone in Hartmann et al, 2018, 281). This trend is occurring in the wider context in which travel, and in particular 'leisure', is seen as 'integral to a good life', whereby individuals are allowed the freedom to express themselves through their choice of destination (Raymen and Smith, 2019, 1).

For John Lennon (2017), the popularity of dark tourism hinges on society's fascination with our ability to do evil. Dark tourism has enabled historical

**Figure 1.1:** Paris morgue, double suicide, 1874

Note: An early form of dark tourism included trips to the morgue. Some morgues offered guided tours, while others allowed people to view certain cases as displayed in this image.

Source: Photo Researchers (nd), 'Paris Morgue, Double Suicide, 1874'

and heritage sites to become popular destinations that, while encouraging consumerism, also educate and inform the public. In essence: 'dark tourism symbolises sites of dissonant heritage, sites of selective silences, sites rendered political and ideological, sites powerfully intertwined with interpretation and meaning, and sites of the imaginary and the imagined' (Stone in Hartmann et al, 2018, 281).

Many dark tourist sites are unquestionably cultural and historic sites of importance. Museums, sites of atrocity and galleries all offer tourists the opportunity to visit cultural destinations that are imbued with meaning. As Reynolds argues, 'dark tourism both informs and is informed by other forms of cultural expression and knowledge production' (cited in Hartmann et al, 2018, 289). That is, dark tourist sites present cultural narratives to tourists and are 'an increasingly pervasive feature in the popular cultural landscape' (Stone, 2009a, 32–33). In the case of dark sites associated with crime and justice systems, these narratives can foster punitive sentiments, at other times question these sentiments, result in indifference (Walby and Piché, 2011) or even evoke entertainment and titillation, much the same way that other leisure activities associated with 'true crime' do.

Dark tourism has become a cultural activity for many people worldwide; as such, cultural criminology offers a robust theoretical lens to examine how sites of crime are portrayed and whether there has been a blurring of lines between what is portrayed in popular culture, and what is 'real'. For Ferrell et al (2008, 2) 'cultural criminology explores the many ways in which cultural forces interweave with the practice of crime and crime control in contemporary society. It emphasises the centrality of meaning, representation, and power in the contested construction of crime'. Drawing upon theoretical frameworks associated with cultural criminology, researchers have successfully demonstrated that dark 'crime' tourism sites are culturally significant to society and are often open to interpretation and 'meaning-making'. Indeed, such sites allow tourists to view 'closed' or restricted spaces in a contrived manner that typically reinforces 'cultural associations and stereotypical images of criminals' (Brown, 2009, 193).

Despite the progress in academic research, dark tourism as a theoretical framework remains 'theoretically fragile, raising more questions than it answers' (Sharpley, 2005, 216). One such question centres around 'what is so "dark" about dark tourism?' (Bowman and Pezzullo, 2010, 188). Or simply, what makes a site a 'dark' site, as opposed to a 'heritage' site? There are also questions surrounding labelling tourists or sites as 'dark', because this makes an implicit claim that 'there is something disturbing, troubling, suspicious, weird, morbid or perverse about them, but what exactly that may be remains elusive and ill-defined because no one has assumed the burden of proving it' (Bowman and Pezzullo, 2010, 190). Linked to this is the notion that some tourist sites have noble purposes such as to educate, to provide a religious or spiritual connection, or as part of a pilgrimage of nationally important cultural sites, while other sites are centred more around 'entertainment' and 'selling' and commodifying darkness. These questions and others will be explored throughout this book, with particular attention to rural sites of dark tourism.

## Development of sites

Death, pain, tragedy, disaster and atrocity permeate our society, yet not all sites associated with these issues become dark tourist destinations. Lennon and Foley (2000) argue that global communication technologies have essentially shaped perceptions about which sites hold significant history, and therefore which sites of death, disaster and atrocity are likely to become popular tourist destinations. In general, it is believed that a site needs to be invested with cultural or political significance to attract visitors.

It is important to note that the term 'dark tourism' is a term that has been applied without the consent of the tourism industry, and the label has been rejected by some tour operators (Light, 2017). There can be significant opposition to such sites, for example, from victims' families, local community members and civic and local authorities (Lennon, 2010). According to

Lennon (2010, 222) 'sites of crime and decisions about their development thus remain emotionally charged and ethically complex. However, they can also be indicative of political and ideological imperatives at a local and national level'. All dark tourist sites face a range of considerations when establishing and running their organisation, including the immediacy and ethics of development, if tourism is purposeful or accidental, whether the focus is on death and suffering, the reputation of the destination and the sites' primary purpose.

*Immediacy and ethics of development:* The establishment of dark tourist sites usually requires the consent and implicit assistance of the community surrounding the location. A major component in establishing community support is the consideration of the timing of the development (Kim and Butler, 2015). In some circumstances, travelling to a site of death directly after the event is acceptable, for example laying flowers at the site of Princess Diana's death (Lennon and Foley, 2000), while travelling to other sites (particularly for leisure as opposed to memorial purposes) only becomes acceptable with a significant lapse of time. Depending on the severity or 'darkness' of the event, a significant amount of time needs to have elapsed before dark tourist attractions are considered 'suitable'. The more time between the death/tragedy/atrocity, the more time the community has to heal and fundamental questions about whether the site is ethical or unethical become less problematic (Kim and Butler, 2015).

While most dark tourist sites are structured and organised exhibitions that display recent or distant historical tragedies, other sites occur more spontaneously and may start without any official tour or organisation. For example, several of the sites throughout this book, including Belanglo and Snowtown, started as spontaneous and unofficial sites that tourists visited out of curiosity or to pay their respects. In both these cases, there has been interest in establishing official 'tours' and associated tourist-information activities around the serial killings. However, in both instances, the community belief that the killings are too recent has inhibited tour operators. In many cases, communities have worked together to prevent dark tourist activities, including demolishing sites where death and atrocity have occurred (Lennon, 2010).

*Purposeful or accidental:* In some instances, organisations or local governments will intentionally, or purposefully, construct sites, attractions and exhibits to create a 'dark tourist' site. In other cases, sites become dark tourist destinations ' "by accident" because of their relationship with turbulent and tragic events' (Stone, 2006, 148). Accidental tourism can in turn lead to purposeful, and more permanent tourism, as in the case of Port Arthur, which will be explored in Chapter 3.

*Focus on death and suffering:* Tour providers will often need to decide on how much attention (and how) they choose to focus on death and suffering.

*Destination reputation:* Particularly in the case of dark tourism activities surrounding criminal acts, local administrators may be reluctant to create tourism industries that promote their town as a place where criminal events have transpired and/or that may denigrate the reputation of their town.

*Primary purpose:* Sites will be developed for a range of reasons, including political reasons (for example, to remember an event or to educate visitors about a particular period of cultural history), entertainment purposes or economic gain (Stone, 2006, 148). The primary purpose of the site will frequently determine the 'type' of site it becomes. According to Lennon (2010), the content, imagery and narrative used to create and present a dark tourism site can raise controversy and therefore questions about the ethics of a site. As such, the type of content displayed (and any associated merchandise) will be carefully considered and aligned with the primary purpose of the site.

In turn, the type of site will be influenced by consumer tastes, media narratives of the site and the original death/tragedy, and the wider political and cultural climate (Seaton, 1999, cited in Stone, 2006, 150). As Lennon (2010, 226) notes, 'the interpretation of artefacts and the exhibition of materials is an inherently ideological practice ... and [the] existence of a site will be threatened if it fails to conform to a political perspective of the past'. The media will also have a role in this selection, as sites often display media coverage of events and thus can shape the types of information displayed to visitors.

## Types of sites

There are many different types of dark tourism sites with varying aims and levels of 'darkness'. In attempting to develop classifications of dark tourist sites, academics have looked at a range of issues, including how the site has been presented and what type of information is available: whether the site promotes remembrance, commemoration, education or simply entertainment; whether the site trivialises events or presents an inauthentic experience; whether empathy is invoked from the visitor; and the 'distance' between the event being portrayed and the actual event itself (this relates to distances in geography and time) (Stone, 2006). 'Black' or the 'darkest' forms of tourism allegedly occur where a site exists to satisfy a fascination with death (Sharpley cited in Stone, 2006).

As such, several academics have proposed varying classifications of dark tourist sites. Dann (1998, cited in Stone, 2006) argued for five principal categories: 'perilous places'; 'houses of horror'; 'fields of fatality'; 'tours of torments' and 'themed Thanatos'. Seaton also developed five categories of travel that were distinct from Dann, including travel to sites featuring public enactments of death; historical sites of individual or mass deaths; memorials or sites such as graveyards, crypts and war memorials; sites providing evidence or representations of death (such as a museum located at a different site containing exhibits such as weapons of death) or exhibits that reconstruct specific events; and sites that provide re-enactments of death, adding a theatrical spectacle to a site were public enactments of death occurred (Stone, 2006).

However, the most comprehensive classification was developed by Stone in 2006 and proposes seven 'dark suppliers' of dark tourism: 'dark fun factories', 'dark exhibitions', 'dark dungeons', 'dark resting places', 'dark shrines', 'dark conflict sites' and 'dark camps of genocide'. Additionally, Stone (2006) proposed a 'darkest–lightest' spectrum which could help to explain the type of experience a dark tourist site offers. The 'darker' sites most commonly have a 'higher political influence and ideology' and are actual sites of death and suffering, are more authentic and are aiming for an educational experience (Stone, 2006, 151). In addition, such sites are typically created 'closer' in time to the event in question. Contrastingly, 'light' sites are those with 'lower political influence and ideology' that are orientated towards providing entertainment or a romanticised view of the death and suffering associated with the site (Stone, 2006, 151). These sites are associated with high levels of tourism infrastructure and sites that represent events that occurred a long time ago. Sites will often move along this spectrum with changes in broader social, cultural and political shifts, with how exhibitions are presented and interpreted, and as more time elapses between the event and the site. Within this book, several of Stone's 'dark suppliers' are investigated, including 'dark conflict' sites (see Chapters 2 and 4), 'dark shrines' (see Chapter 6), 'dark exhibitions' (see Chapter 7), 'dark dungeons' (see Chapters 3, 5 and 7), 'dark fun factories' (see Chapters 3, 4, 5 and 7), 'dark resting places' (see Chapter 7), as well as another category, 'atrocity and catastrophe' (see Chapter 7).

## Presentation of sites

Most, if not all, museums and dark tourist sites have contested histories and understandings that visitors in essence 'witness' or consume. Part of the role that tourist site managers take on is determining which stories or narratives to present to visitors. As cultural studies academic Jacqueline Z. Wilson notes:

> An abiding characteristic of historical tourist sites is that those whose task it is to research, compile and (re)present the aspects that comprise the sites' 'history' tend to come under a variety of pressures to 'choose' one, or a relatively narrow category, of narratives from among the complex of individual accounts that inhere within the sites. This process can result in what may be termed 'frozen' narratives – the effective denial of the sites' alternative stories and the consequent

negation of the individuals whose voices carry those stories. (Wilson, 2011a, 562)

Dark tourism sites usually offer a very specific, and cultivated, interpretation of a particular historical narrative. Museums, exhibitions, dungeons, fun factories and so on tend to present a narrative around an 'event-based view of violence, leading the tourist through the history and details of a specific tragedy' (Robb, 2009, 54–55). As such, curators of exhibits and sites as a whole, participate in an 'unrecognised ethical dilemma' of determining 'whose history' prevails in being presented (Lennon and Foley, 2000, 162).

The physical space of a site regulates visitors' conduct and understanding of a site (Bennett, 1995, cited in Welch, 2013). Whether the physical space consists mainly of architecture or vegetation, it plays a vital role in shaping how a visitor understands and interprets a site. Architecture, particularly for prison museums, can express 'order' through its 'sheer size, scale and symmetrical design' or as a means of providing authentic proof of a historical period (Welch, 2013, 482). Architecture and physical barriers can be employed to encourage visitors to enter an exhibit at a certain place and then proceed through the site in some organised manner, thus further influencing the narrative that a tourist is exposed to. Even in natural settings, the physical space of a site can have a deep and powerful impact on the visitor.

Exhibits within a site are carefully selected and displayed to attract visitors via advertising and promotional activities (Welch, 2013). Objects need to be significant, rare or revelatory – or in essence, they need to be able to produce meaning for the tourist (Welch, 2013). Exhibits, accompanied by labels, seek 'to render violence as explainable and knowable through education and information' (Robb, 2009, 55). As part of this process, labels provide order and meaning to something that may otherwise be incomprehensible to most visitors, or even 'render it aesthetically pleasing or even beautiful' (Robb, 2009, 55). Some objects will hold 'sinister appeal' because it is assumed (or the visitor is told) that the object was used to commit horrible acts, such as hanging scaffolds (Welch, 2013). However, some sites offer minimal interpretive information – adopting the strategy of minimal exhibits and textual information – thus encouraging the viewer to form their own views or have a more primal connection to a site.

The authenticity of a site is viewed as an important element for dark tourist sites, particularly if they are seen as being on the 'darker' side of Stone's scale. To achieve authenticity (or at least the perception of authenticity) sites often display original objects, such as weapons, torture devices, newspaper or documentary evidence from the time, to increase the feeling of an authentic experience for tourists (Robb, 2009). Prison museums, by virtue of being located in a former prison 'are rightfully judged to be authentic' (Welch, 2013, 482). However, at times, this can create obstacles for some sites, such as where demolished or deteriorated sites may need repair. These repairs can be construed as being inauthentic and therefore detract from the perceived authenticity of the site itself. Dark tourist sites, like museums, rely 'on the mutually reinforcing relationship among objects, images, and space: altogether linking to a particular site' (Welch, 2013, 483).

Most dark tourist sites will have aims and philosophies guiding their exhibits and tours. As noted earlier, these aims will vary differently across tour providers depending on what type of site it is, what customers they want to attract and the types of messages that are conveyed to visitors. Regardless, the artefacts and exhibits within these sites present a complex narrative(s) often utilising texts, histories, imagery and interactive activities. While there is no one 'correct' reading of these exhibits or tours – and visitors are likely to walk away with a wide range of emotions and understandings of the site they have visited – it is still true that such sites offer some form of historical record (Lennon, 2017). Sites will be presenting particular narratives, and while the interpretation of this can be varied, an analysis (or deconstruction) of the content being presented to visitors still yields vital information about the political and cultural motivations of sites.

## Motivations and reactions

Another concern with dark tourism literature is that the motivations of tourists are often overlooked or generalised. Further, the label 'dark tourist' often has negative connotations that 'is somehow deficient in requisite morals, historic comprehension, and cultural codes, and who possesses an innate inability to be elucidated by memorial messages' (Stone in Hartmann et al, 2018, 291). In reality, tourists' motivations for visiting sites will vary, as will their reactions to the sites themselves. For example, previous literature has suggested that tourists can be motivated by idle or morbid curiosity, intellectual interest, reverent pilgrimage (Wilson, 2011a, 562), the desire for 'ghoulish titillation' or to be entertained (Wilson, 2004, 13), moral obligation or leisure (Golańska, 2015, 778), to confront death and reconnect with ancestral land or victimised forebears (Kidron, 2013), 'curiosity about the unusual', 'attraction to horror' and/or 'a desire for empathy or identification with the victims of atrocity' (Light, 2017, 285), to enhance national pride (Tinson et al, 2015, 875) and/or to learn about past atrocities in the hope of preventing future atrocities (Robb, 2009). The available research indicates that the most common motive is to learn and understand historical events (Light, 2017). Due to the varied motivations associated with travel to dark tourism sites, Stone (in Hartmann et al, 2018, 292) argued that there is no such thing as a 'dark tourist' – only people that are interested and 'engaged in the historic and social reality of their life-worlds', and as such, engaging in dark tourist activities does not make someone 'deviant' (Stone, 2017).

Motivations and reactions are strongly linked. Take, for example, the following section from Heather Brook describing her visit to the Adelaide Gaol in Australia:

> In the 'New Building', children are zooming in and out of cells, making their own cops and robbers stories come alive. Me, I can't cross the threshold. I tell myself that I ought to step into a cell; I remind myself that it's safe. But all I'm aware of is that this is not my place. This is a place marked by the profound historical absence of people like me, people whose brothers (or sons, fathers, husbands, friends) loved and missed them. I don't believe that ghosts haunt the Adelaide Gaol. But I can see that grief, despair, and misery coat these cells over and over, peeling away like thin institutional paint. (Brook, 2009, 267–268)

For those without a personal connection to the site, there may be little hesitation to cross the threshold of a cell – and for those who are younger, decommissioned prisons provide the perfect cops and robbers play setting; yet for those with a deep connection to a site, the reaction to visiting that site is likely to be very different, as illustrated by Brook.

As such, there is a further element of Stone's dark or light scale at work here. Dark tourist sites offer a range of activities that, to some, are distasteful and disrespectful. Even the simple act of eating a packed lunch at one of these sites can be seen as objectionable behaviour, particularly to those who have been personally touched by such tragedy. Other sites have now become function centres, offering consumers the opportunity to host parties or even weddings at these venues. For those with a personal connection to the site, this can be an incomprehensible development:

> The Adelaide Gaol is not Auschwitz. But it, too, is a place of horror and tragedy. It's also a function centre. I look at the Law Students' poster again. How can a group of law students celebrate passing their exams in this place where so many lawyers' clients despaired and died? When the Royal Commission into Aboriginal Deaths in Custody recommended the establishment and support of Indigenous tourism, this doubly 'dark' tourism is surely not what it had in mind. (Brook, 2009, 268)

Reactions to sites differ depending on the motivations for visiting, personal connections to the site and the level of understanding one has for the site. As Robb (2009, 52) has argued, experiences and reactions at dark tourism sites are not uniform or objective – they are 'subjective and extremely individual'.

Most dark tourism research on motivations for travel tends to focus on adults. However, as Israfilova and Khoo-Lattimore (2019) argue, the motivations and reflections of children are also important for those who plan and manage exhibits. Many dark tourist sites offer school excursions and activities designed specifically for children (whether as part of a school tour or in the form of school holiday activities), and as shown through the example Brook provides, many children attend these sites as part of family trips. For children, such sites can provide a powerful 'history-telling role … [that] enables children to understand the true history of their past, build an identity and be confident about their knowledge of their ancestors' (Israfilova and Khoo-Lattimore, 2019, 480). Yet, the reactions of children to such sites require careful monitoring and awareness. Children may struggle with images and narratives of death, suffering and violence (Israfilova and Khoo-Lattimore, 2019).

Regardless of the motivation, visitations to such sites involve some form of leisure activity (Golańska, 2015) that encourages sites to create experiences and infrastructure similar to entertainment parks to encourage visitors to attend. Infrastructure and 'activities' can be important marketing tools, particularly for those sites located in rural and remote areas (to be discussed later in this chapter).

## Tours

Many dark tourist sites offer tourists organised tour activities as part of their tourism infrastructure. There is a variety of different types of tours that again reflect where a site will sit on Stone's 'dark–light' scale of tourist providers. A common strategy adopted by sites is to employ actors to 'perform' for tourists. These performances may be part of an organised tour, or they may simply occur at the site where tourists can watch if they want to. When witnessing such performances, Bowman and Pezzullo (2010) argue that tourists are audiences of multimedia and oral-dramatic events rather than consumers of a product. Within these performances, the tourist can sometimes be offered the opportunity to become co-performers or be fully immersed in the performance by becoming a prisoner or a guard at a police watch house for example.

Some dark tourist sites employ 'fright tourism' strategies. For Bristow and Newman (2004, 215), 'fright tourism' refers to tours that offer individuals 'a thrill or shock from the experience'. Brown (2009, 91) argues that the presentation of penal history as a form of trauma at prison museums is 'inevitably embedded in spectacle and thrill-seeking'. Similarly, ghost tours are often very successful marketing techniques that encourage tourists to experience a 'different' side of a site. Yet, as Staples (1995, 36) argues, 'they sell nothing'. Tour guides are very clear that a ghost is unlikely to appear – yet it adds further anxiety to the experience.

A growing number of sites and tour companies are offering tourists the opportunity to 'recreate "last journeys" in space and time' which 'raise questions of both taste and sanctity' (Lennon and Foley, 2000, 164). For example, on the *Helter Skelter* tour in Los Angeles, tourists were allowed to wash their hands at the same tap that the killers washed their hands after murdering Sharon Tate (Podoshen et al, 2015). As such, tourists become part of the scene themselves.

## *Promotional material, gift shops and souvenirs*

Sites of death and disaster attract visitors regardless of whether there is a formal tour operating. People discover sites through promotional material such as websites, pamphlets, posters, blogs, social media, documentaries, newspaper articles and information centres, as well as word-of-mouth. Much of this promotional material 'plays a key structuring role in tourism' by enabling the reader 'to journey vicariously in the words and images of the travel narrative' (Robb, 2009, 53). Brochures and other promotional material also play a role in providing visitors with information on how to behave at a site, the range of activities provided and as an official souvenir of a site.

Souvenirs offer an important opportunity for tourists to collect 'objects that mediate memories' of their tourist experience (Cave and Buds, 2018, 708). According to Gordon (1986), almost everyone collects souvenirs to be reminded of special moments and events. These souvenirs are then gifted to loved ones, or placed within the home, serving as a talking point of their travels or a visual and emotional reminder of their experiences. Acknowledging this, many dark tourism sites, like other heritage sites such as museums, offer tourists the opportunity to purchase a memento of their visit.

Gift shops can offer a crucial source of revenue for heritage sites (Brown, 2013), yet these sites, particularly dark tourist sites, are often criticised for having a shop. Sites are accused of being commercial and placing priority on economic imperatives rather than focusing on providing an educational or cultural service to the community. This issue is intensified when considering dark tourism sites. As Brown (2013, 273) notes 'shops in dark tourism institutions are highly constrained by issues of taste and decency' and these sites and their gift shops can be seen to be highly provocative (Stone, 2009b).

Returning to Stone's light and dark scale, the darkness of a site is likely to affect the type of souvenir on offer. Merchandise is a way to 'reconfirm the institutional mission' and the 'employment of obviously profiteering behaviour and distracting commercial practices would serve to undermine their educational and memorialising identity' (Brown, 2013, 278). Further, Brown argues that tourists are unlikely to buy frivolous merchandise from darker sites. As such, sites need to balance their commercial needs with concepts of decency and taste to remain profitable and true to their purpose.

There are many different types of souvenirs. For example, many sites offer generic souvenir items such as art and jewellery that 'are chosen to essentialise, universalise, and offer objectivity and to symbolise the realities of events that took place. They also serve to distance the consumer from the deepest emotions' (Cave and Buds, 2018, 722). Others offer a range of 'child-friendly' souvenirs such as children's clothing, plush toys and magnets. This serves not only the family tourists but also adults who like to take something home to children (Gordon, 1986). Many offer educational books on the site or surrounding narratives. Gordon (1986) argues that there are five distinct types of souvenirs. The first is pictorial images, usually in the form of postcards, but also books or photos that tourists themselves take at the site.

Second, 'piece-of-the-rock' souvenirs are natural objects or materials taken from the environment. Sometimes this can be sold in a gift shop, but at other times, tourists (illegally in many cases) take items from the site itself. For example, following the media frenzy over the Jeffrey Dahmer serial killings in Milwaukee in the United States, tourists from as far away as Japan began to visit the site of his crimes (Fox and Levin, 1994, cited in Gibson, 2006). The property owners hired guards and surrounded the property with barbed wire to deter scavengers. This in turn led to some visitors offering to pay up to US$100 for a 'souvenir brick'.

Third, symbolic shorthand souvenirs are objects that represent the site itself, such as a plush convict puppet magnet. Fourth, a 'marker', such as a printed T-shirt bearing the name of the site. And finally, local product souvenirs such as local crafts or food are on sale, which importantly supports the community.

## Creating identity

While the main intent behind dark tourist sites is to attract (paying) visitors, places of death and suffering also play an essential role in creating, crafting and presenting cultural and national politics of identity. Many dark tourist sites have a significant historical value that is embedded within broader politics of remembrance (Seaton, 2009a, cited in Light, 2017, 284). As such, these sites play an important political role as 'places of collective/national memory' (Light, 2017, 284). As museum scholars Piché and Walby aptly state:

> Museums influence, and are influenced by, public memory and culture. … Memories are formed and forged within museums, which informs how beliefs and norms are conceived of in the present and future, influencing views of self and others. … Visits to museums and other tourism activities are opportunities for meaning making, as well as the

imposition of significance on historical objects and events. (Piché and Walby, 2018, 567)

While Piché and Walby (2018) are discussing museums specifically, the sentiments apply to most dark tourist sites and the role of sites in creating 'meaning-making' and 'collective memory' within Australia, particularly in relation to the idealisation of convicts and bushrangers. However, the power of sites to imbue place identity can also be quite problematic, particularly for rural areas where some sites of death and atrocity have been reinvented as dark tourist sites and, as such, gained popularity and notoriety for something that is possibly painful and/or shameful, when the community would rather distance themselves from that dark and painful past.

Sites may be initially created telling one narrative and then later converted to tell a different version of the same events. Take for example the site of Port Arthur, Tasmania. At the time of its closure as a convict settlement, residents of the town tried to distance themselves from dark tourist activity, yet they were unsuccessful – tourists continued to travel to the remote area to 'experience' a piece of history. As social and political attitudes towards convicts changed (that is, away from the notion of the 'convict stain', which will be explored in Chapter 3), the site of Port Arthur became celebrated. As such, according to Wilson (2011b, 203), 'sites often pass through phases of contested meaning or significance, being reviled as shameful in one era, only later, as historical sensibilities shift, to be celebrated'. Despite these shifts in meaning, the site can often continue to play a significant role in the process of 'nation-building'. Sites that embody the 'darkest narratives' affirm social memory and national identity and encourage both official and unofficial acts of remembrance from visitors (Wilson, 2011b, 203).

Many dark tourist sites around the world are popular school group tour destinations. Excursions to sites of death, tragedy and atrocity, especially when it relates to national identity, are powerful educational tools. Through participating in immersive excursions, students can explore the environment and the 'reality' that is being presented. According to Wilson:

Each year thousands of Australia's primary and secondary schoolchildren visit such environments. The excursions they undertake can range from a short walk to a nearby cemetery or war monument and back, to a full day of touring, say, a former prison or convict site, or an extended period visiting interstate memorials or even overseas battlefields. (Wilson, 2011b, 203)

Although such sites are often used to investigate and reaffirm, notions of cultural identity, the fact remains that the interpretation of these sites will still be contested – there is no one 'right' way to interpret such sites. In addition, essential information is often excluded from such sites. Wilson (2011b) has recognised the overall lack of information on Aboriginal incarceration in many Australian penal museums. In regards to school tours then, teachers may need to invest in significant additional learning material to make students aware of the contested information present in many of these sites.

*Ethical considerations*

In the previous discussions, ethics has been a common theme of dark tourism. As will be demonstrated throughout this book, whether a dark tourism site is deemed 'ethical' or not will depend upon the time after the event in which tourism begins, the way tourism is conducted, what the tourism is 'promoting' in terms of education or entertainment, and the potential sale of souvenirs. For example, as Chapter 6 highlights, tourism that is proposed 'too soon' after a tragic event is likely to be perceived as unethical by the local community, and therefore fail (although not always, as will be seen in Chapter 3). In the case of the serial killings at Snowtown and Belanglo State Forest, the local communities fought strongly against tourism activity, arguing that it was unethical to profit from the death and suffering of others. If the tourism presented is more sombre and reverential (compared to 'fright' tourism), it is generally viewed as more 'ethical', as are sites that promote education over entertainment.

There are also issues with presenting the 'prevalent' version of history (or even an authentic account) when there are divergent accounts and the ethics associated with providing interpretations to visitors (Lennon and Foley, 2000), or the ethics of presenting images of suffering or instruments of torture (Lennon, 2018). Finally, the sale of souvenirs can also be seen as unethical. Referring to Snowtown again, one entrepreneurial resident was 'kicked out of town' for selling kitsch souvenirs too close to the discovery of the bodies (Kim and Butler, 2015, 86) because it was deemed unethical.

# Dark tourism in Australia

While the dark tourism industry has boomed internationally, there have been some questions about the scale and availability of dark tourism sites within Australia (Wilson, 2008a). In particular, questions have been raised over the lack of sites of war, genocide and wide-scale atrocity.

Prison tourism is seen as the main form of 'dark tourism' available within Australia. According to Walby and Piché (2011, 452) 'visiting a decommissioned prison or jail turned museum is now a common form of tourism and leisure', and in the Australian context, 'penal museums ... are particularly popular cultural institutions, not only among tourists but locals with whom the museums resonate as part of a mythologized convict past' (Walby and Piché, 2011, 454). Wilson (2011a) claims that at least one in ten, but possibly one in six, of today's Anglo Australians are descended from convict transportees. In addition, Australia's 'convict origins' remain a dominant part of national cultural memory and identity (Wilson, 2011a). As such, many (local) tourists to convict and penal sites will offer 'some degree of personal connection between their own sense of national identity and the notion of imprisonment' (Wilson, 2011a, 564).

It is important to keep in mind that 'Australia's unusual historical relationship with both the imprisoned among its populace and the buildings in which they were housed means that questions of narrative "choice", and concomitant narrative exclusion, are of especial moment' (Wilson, 2011a, 563). That is, curators need to be particularly aware of the narratives they are promoting and cultivating. There is a tendency 'of "othering" the imagined inmates of the prisons visited' (Wilson, 2008b, 331).

While this book recognises the importance of penal tourism in both urban and rural areas of Australia, dark tourism within Australia encompasses a lot more than penal museums, particularly in the rural context where Australia's historical events open up unique and, at times, overlooked dark tourism activity within Australia. As Staples (1995, 35) notes, tourists and, in particular international tourists, are attracted to the convicts and prisons, as well as bushrangers. There is also tourism around colonial violence, crime-related violence and death, and more recently, there has been some speculation that natural disaster sites have the potential to become a part of the 'dark tourism' industry. Many of these sites are located in rural and regional parts of Australia and will be explored throughout this book.

## Rural dark tourism

Where dark tourism in Australia has been examined, it has generally been focused on those penal sites within urban settings, for example, Old Melbourne Gaol in Victoria, Fremantle Prison in Western Australia and so on. There has been some analysis of regional carceral sites such as Old Dubbo Gaol in New South Wales (NSW) and Port Arthur in Tasmania. And fewer still of the more, 'rural' sites such as Trial Bay Gaol in NSW. Looking more broadly at dark tourism destinations in Australia, there has been some analysis of rural sites such as Snowtown and sites associated with Ned Kelly. However, these studies are few and far between.

Internationally, researchers have recognised the occurrence of dark tourism activities in rural and regional areas, but analysis of these sites has been limited. For example, Lennon (2010) discusses the case of the town of Soham in Cambridgeshire, United Kingdom, which came to receive public attention in 2001 when Holly Wells and Jessica Chapman were kidnapped. What

Seaton and Lennon (2004) term 'voyeuristic visitors' allegedly travelled to Soham – some to leave flowers or children's toys at the site of the kidnapping, others to 'simply observe this ordinary town in the turmoil following the abduction and murder of these two children' (Lennon, 2010, 219). At the time, there were reports that tour buses were rerouted to travel to Soham to facilitate dark tourism. While there is recognition of the travel to 'out of the way' locations to visit such sites, there has been little analysis as to why tourists make the effort to travel or on the impacts on the town itself.

The importance of rural should be more prominent in dark tourism research, particularly research that focuses on penal establishments. Many penal institutions were purposively built in more rural and remote areas, or at least outside the city limits 'perched on hills as fearful beacons for the outlaws at large in the community' (Shehata et al, 2018, 1). Others were built away from communities or other prisoners to isolate inmates. For example, the site for the Point Puer Boys' Prison at Port Arthur was 'selected as a compromise between geographical isolation and administrative efficiency' (Jackman, 2001, 8). The boys were intentionally geographically isolated from the adult male convicts being held at Port Arthur. As such, many rural and regional areas of Australia house decommissioned penal or carceral architecture that has the potential to be converted into dark tourism destinations.

Overall, there has been an oversight of dark tourism activities in rural and regional areas, both within Australia and internationally. This has meant that the locational context of sites and how cultural geography has shaped dark tourist sites have also been overlooked. This is not unique to dark tourism research. As Scott and Hogg (2015) note, most research during the nineteenth century was preoccupied with social and crime problems of urban areas; and rural communities tended to be idealised and ignored from analysis. Further, policy responses to crime and punishment are normally urban-centric and are often inappropriate for rural/regional areas.

## Rural idyll and rural dystopias

When we think of rural spaces, we often conjure an image of beautiful bushland or widespread, peaceful, idyllic countryside. This then becomes the rural idyll or rural utopia and can play a large role in marketing campaigns to attract local and international tourists (Rofe, 2013). Within the rural idyll, communities are portrayed as having wholesome values, strong, close-knit communities and kinship, and a 'landscape of religious piety and social stability' (Rofe, 2013, 270). In the past, the rural has been used to showcase the Australian nation to the rest of the world – the beautiful, romantic outback and bush were promoted as a symbol of nationalism (Beeton, 2004).

However, this rural idyll was challenged and subverted throughout the 1970s and 1980s within popular culture, particularly in films (Scott and

Biron, 2010). For example, the film *Picnic at Hanging Rock* released in 1975 juxtaposes the Australian outback as both beautiful and alluring, but mysterious and dangerous – it is a landscape that is capable of transfixing young girls and luring them into an environment from which they would never be seen again. In another example, the bushranger Ned Kelly has long been a symbol of Australian heritage (explored further in Chapter 4), and a large part of this cultural story occurred within 'rural Australia'. The rural landscape thus remains a part of national marketing campaigns and the national identity whenever Ned Kelly is featured. Yet, the idyllic and beautiful landscape that is presented, like in *Picnic at Hanging Rock*, becomes subverted to become a place of conflict, hardship and violence (and also horror). Rofe refers to this as rural dystopia:

> the idea of rural dystopia is an alternate and darker reflection of the discourse of the rural idyll. Where rural utopias reflect community cohesion, harmony with nature and physical and moral vigour borne of honest labour, rural dystopias imply inbreeding, struggle and a sense of callous indifference borne of hardship. (Rofe, 2013, 263)

More recently, media reports have tended to label some rural areas as 'evil', or with 'hidden rot' in previously idealistic spaces (Rosser, 2013, 76). The rural has become dystopian in these locations. Away from the 'civilising qualities of the urban', the rural dystopia is a marginal space of disharmony, conflict and trauma (Rofe, 2013, 269). Dystopia can also represent an indication of the decline in society and create a fear of the downward trajectory of the world (Podoshen et al, 2015). As such, it can act as a catalyst to change our ways and thus improve our future. For some, travelling to such sites of dystopia enables tourists to 'engage in making sense of the process, the aesthetics and the emotion of dystopia' (Podoshen et al, 2015, 325), or rather to allow tourists to reflect on 'what went wrong' in the rural idyll. Podoshen et al (2015, 325) coin this type of travel as 'dystopian dark tourism' and argue that 'engaging in these tourism experiences reduces fear and insecurity about death and dystopia'. While dystopian dark tourism is not limited to rural and regional areas, such areas provide a starker contrast to the idyll.

In many cases, there is a combination of the rural idyll and rural dystopia. The beauty of the landscape is juxtaposed with the site which inevitably becomes the rural dystopia. Take, for example, Trial Bay Gaol or Port Arthur Heritage Site – both sites represent 'dark' encounters which are starkly set within a beautiful and tranquil setting (see Chapters 3 and 5 for further details). Examining, or 'reading the landscape' as Fagence (2017, 453) has suggested, is vital to understanding the whole tourist experience and how rural and regional sites offer a different experience within dark tourism.

Ruins (or sites) are used to 'playfully conjure up the horrors of the past' while, at the same time, the landscape acts to reduce the threatening nature (Staples, 1995, 36) of many of these sites.

Rural dystopias offer communities an opportunity to invest in commercially viable dark tourism activities. Some communities have been stigmatised by the media, while others openly embrace 'darker' tourism activities. Regardless of how dark tourism sites in rural areas originate, the role of the rural idyll or rural dystopia that is portrayed at each of the sites, how this affects the message of the site (education as well as entertainment) and also how it affects the local community is an important area to understand. There has been very little analysis of how geographical location plays a role in dark tourism sites, or how the surrounding communities and their history are portrayed. There is also a lack of analysis of how sites may adopt strategies to make their site look more 'idyll' or more 'dystopian' and how this will change where a site sits on Stone's (2006) 'dark or light' site scale, the types of activities it offers and the type of tourist it is appealing to.

## Distinguishing aspects of rural dark tourism

Sites with significant cultural attraction may increase tourist activity, with previous studies suggesting that between 35 per cent and 70 per cent of all visitor tourist activity involves a 'cultural tourism element' (McKercher, 2001, 31). While these figures suggest that cultural tourism may be a significant 'drawcard' for tourists, the same research discovered that cultural tourism was not the primary motivation for travel. Instead, cultural tourism was seen as the secondary or tertiary motive for a tourist to initially travel to a location. As such, while cultural attractions appeal to tourists, they are rarely the motivating factor for the visitor going to that location.

There are many distinguishing aspects of rural and regional dark tourism travel. For example, 'rural tourism in Australia is primarily a domestic tourism activity, in many cases with over 90 per cent of the visitors to non-urban areas being Australian' (Beeton, 2004, 126). These sites are often hard to travel to and are less likely to be visited just because someone is in the location (as opposed to, say, the Old Melbourne Gaol, where a visitor to Melbourne may go to the site as 'something to do' rather than the main purpose of their travel). According to Tinson et al (2015), many visitors do not specifically travel to dark tourism sites to see the macabre. Instead, they are in the area and are looking for activities to engage with that relate to culture and education. As such, 'a remote site needs more variety of function to justify the visitor's travel time, while a site in a capital city may seek to hold a visitor for 90 to 120 minutes and maximise the throughput' (Staples, 1995, 36).

There are numerous competing demands that rural/remote communities can face when considering adopting a dark tourist strategy (or having the strategy forced upon them), and these factors will shape how tourist providers and communities of sites develop strategies on how to implement dark tourist sites, or how to manage the balance between education/entertainment/ethical practices. While many of these issues overlap with those outlined earlier, for example, rural dark tourist locations can be vital for shaping regional and national identity (Piché and Walby, 2018); there are unique considerations that are investigated throughout the case studies.

Many Australian communities are seeking sustainable economic growth in areas outside of their traditional areas (for example, agriculture, mining and so on), and often look to tourism as a supplementary source of community income (Walmsley, 2003). This can be especially the case where there are distinctive features, landmarks or marketing opportunities (Pearce et al, 2003), such as with bushranger legends in Australia. According to Rofe (2013, 262), 'consumption-oriented economies are perceived as the panacea for rural decline. Tourism is a key element in the new rural economy as rural places are re-imagined, re-packaged and re-presented for a predominantly urban market'. At times, the rural idyll is marketed, and at other times the rural dystopia is. Where tourism boosts the economy of a local area it is usually seen as enhancing the social fabric of a community (McKercher, 2001).

In many regional communities, local governments will become involved in the creation and operations of cultural and heritage tourist attractions (McKercher, 2001). According to tourism scholar McKercher, there are several reasons for this:

> They do so partly for altruistic reasons in the belief that the attraction provides broader community benefits beyond tourism. They also do it for pragmatic reasons: many of these attractions are not commercially viable and, therefore, require an ongoing taxpayer subsidy to retain them. (McKercher, 2001, 29)

Local governments can also act as an intermediary for developing 'sensitive solutions to the demands of dark tourists that are neither offensive to residents nor indeed, the families of the victims' (Kim and Butler, 2015, 87). Regardless of the reason for government intervention, the impact of such intervention is that the narrative presented can be influenced by political values and may be a source of controversy for an area, especially, McKercher (2001) argues, if the level of subsidisation is viewed as excessive and outweighs benefits to the community.

Any form of tourism, dark tourism included, requires some form of infrastructure to thrive and contribute to building a community's economy

(Kim and Butler, 2015). While some locations become informal dark tourism sites, such as memorials at serial killing sites, many sites are carefully planned and catered for. In the rural and regional context, it is vital that a community can provide sufficient infrastructure to sustain and encourage long-distance travellers. For example, accommodation is likely needed in rural and regional areas to enable overnight trips for long journeys. Sufficient signage or acknowledgement of the site is required and other tourist-type activities in the community or at a nearby location are desirable (Kim and Butler, 2015).

Theme-linked travel can provide multiple rural and regional communities with combined infrastructure that incentivises tourists to travel long distances to participate in narratives about a particular person or cultural history. For example, in Tasmania, multiple convict heritage sites in close proximity encourage tourists to travel to Tasmania for an extended time and explore as many locations as possible. Similarly, other areas can create a 'trailscape' that is:

a specially contrived mode of story-telling which derives its significance and distinctiveness from its capacity to tell a story using both action-relatedness and place-relatedness as they are spread across a range of sites which are connected principally by the details of the story, and which have both chronological and geographical associations, and for which movement between the various sites is facilitated by a convenient trail. (Fagence, 2017, 455–466)

Touring rural and regional parts of Victoria to follow Ned Kelly's original trails provides a perfect example of this. Communities can work together to create a combined tourist experience that makes the trip feasible in terms of time and money for the tourist, while still benefiting several communities' economies.

Tourism to rural areas is generally seen as a benefit for a community – it can improve the economy, increase recreational facilities and cultural activities and enhance a sense of community pride in residents (McKercher, 2001). Heritage tourism can ignite an enthusiasm for connecting tourists with their culture and history, and indeed many convict sites, for example, offer tourists the opportunity to explore the convict records to see if they have a distant convict relative.

Yet, what happens when a town feels that a site detracts from community pride and would prefer to distance itself from painful and dark moments in its history? At other times, rural communities may express antipathy towards tourism where the tour operators are seen as outsiders to the community, or the tours encourage more outsiders to visit the community (McKercher, 2001) for the 'wrong' reasons. There are several examples of rural and regional communities rejecting dark tourism activities. Two Australian examples,

Snowtown and Belanglo will be explored in Chapter 6, however, there are also several international examples.

In America, famous body snatcher and murderer Edward Gein's farmhouse in Plainfield became an impromptu tourist site that 'hordes of curiosity seekers' visited (Schechter and Everitt, 2006, 285), and one company even proposed turning the farmhouse into a 'House of Horrors' tourist attraction (Sutton, 2020). According to Gibson (2006, 54), 'approximately twenty thousand sightseers invaded' the town on the afternoon of one of the pre-auction inspections on 23 March 1958, and the farmhouse was the site for illicit college parties, to the extent that fulltime guards were employed to guard the site. The opposition to these tourist activities was so extreme that the farmhouse was 'torched' by 'outraged townspeople' removing the possibility of turning the house into a permanent attraction (Schechter and Everitt, 2006, 285). Despite this, the location of the farmhouse is still a tourist site (even if unofficial), and tourists continue to post information online about their trips to the site.

While not strictly a rural or even truly regional area, similar events transpired in Gloucester, England when the residence of Fred and Rosemary West became identified. Tourists again travelled to this site and, consequently, the house was demolished after local policy makers consulted with residents. Further, the debris from the house was 'buried some 25 metres underground to prevent souvenir hunters from pillaging the site' (Lennon, 2017, 220). The street was renamed to further deter visitors.

Whether a community accepts such dark tourism activities often depends upon the length of time between the act of atrocity and the establishment of tourism activities. Yet, in many rural and regional areas, dark tourist activities occur regardless of the community's desires. All of these aspects inevitably change the site and the tours and activities on offer. This will be a common theme explored throughout the book.

PART I

# Australia's Colonial
# Tourism Destinations

2

# Colonial Violence

Dark tourism surrounding colonial violence against First Nations populations differs substantially from Australia's more common dark tourist destinations, such as prisons. First, there is less acknowledgement of such sites, second, sites are less developed and 'commodified' and third, there has been a historical focus on silencing Aboriginal narratives of colonial violence, which has made the development of such sites 'hidden' or neglected. Despite the lack of knowledge (or recognition) of such sites, it is clear that the violent dispossession of Australia's Aboriginal population has led to some 'dark tourism' activities, particularly in rural and regional parts of Australia.

For the most part, dark tourism relating to Aboriginal people and the colonial invasion relates to 'warfare tourism', which includes (but is not limited to) travel to battlefields, war memorials, cemeteries and peace parks that may include areas where human remains are located, sites of incarceration and enslavement and locations of frontier violence (Lemelin et al, 2013). In many instances, Aboriginal perspectives of these events have been missing from tourism destinations, and Aboriginal people have been trapped in 'a sort of tourized confinement in the suffocating straight-jacket of enslaving external conceptions. They are caught in the objectifying slant of "Whites", "Westerners" and "Wanderers-from-afar" in an anonymous but continuing process of subjugation' (Hollinshead, 1992, 44). This absence of Aboriginal voices, and a rigid colonial narrative, results in 'selective, partial, biased and distorted' storytelling within sites of colonial violence (Lemelin et al, 2013, 258).

Further, as Wilson (2011b, 211) recognises, 'although sites of Aboriginal suffering abound on Australian soil, very few as yet are widely known'. While these sites may be less well known, tourism does occur, and this is not a recent phenomenon, for example, tourists have been visiting the boab trees since the nineteenth century (to be discussed further in this chapter). For Stone (2006), such locations are categorised as 'dark conflict sites' which 'essentially have an educational and commemorative focus, are history-centric and are originally nonpurposeful in the dark tourism context' (Stone, 2006, 156).

Colonial violence sites within Australia are less frequented by tourists, and perhaps the main way that tourists are exposed to colonial violence narratives (through tourism activities) is via travel to museums and art galleries. Many local, state and national museums and art galleries have been showcasing colonial violence for decades in much the same way that prison museums showcase past penal practices. Historically, many of these accounts have perpetuated 'Eurocentric accounts' which assert the colonial view as the 'proper one' (Tucker and Akama, 2009, 10). As such, some examples of these tourist sites are examined in this chapter to understand how colonial violence is represented and dealt with, with a particular focus on rurality.

## Colonial violence: warfare tourism

The illegal removal (and killing) of Aboriginal Australians occurred across Australia, yet, a large proportion occurred in rural areas, and much rural injustice can be traced back to colonialisation. As Jobes et al (2001) state, law enforcement was often harshly administered in rural/regional areas and ongoing conflicts inflicted through the process of removing Aboriginal people from their land resulted in rigid and brutal control measurements. Further, 'the most violent confrontations against Indigenous people by the colonising populations occurred in rural agrarian areas' (Jobes et al, 2001, 4).

Australia's legacy of colonial violence has been referred to as the 'Great Australian Silence', where talk of frontier violence and war has been suppressed or treated as 'absurd' (McKenna, 2002, 30, 32). In most narratives (until recently), the colonists have been portrayed as the heroes, and the Aboriginal population as barbaric and underhanded (Lemelin et al, 2013). There was also a constant push to colonise the land fully, removing all traces of Aboriginal practices and traces of conflict. For Birch (1999, 63), some places within rural and regional Australia underwent a memorial transformation in the twentieth century that displayed 'confident and authoritative colonial history' that celebrated the pioneers. Such monuments and memorials were constructed in a way that failed to recognise the 'histories and attachment to land of Indigenous people or the violence of the attempted dispossession of these people' (Birch, 1999, 63).

Birch (1999) recounts several 'towns' in rural Victoria that had been abandoned. Despite failing to prosper, those towns were memorialised through signage and other markers, including a giant koala at Dadswell's Bridge to ensure that the colonial successes were immortalised. For Birch (1999, 71), 'the pervasiveness of a colonial history project centred on commemoration and tourism can destroy the ability to remember at all. These sites refuse to move beyond the veil of *terra nullius*, reflecting a colonial past rather than accepting the possibility of a post-colonial future'. Over the past few decades, there has been growing interest in providing a more nuanced and balanced narrative to Australia's history in general, and more specifically, to many of these sites of conflict. First Nations people are being consulted and are participating in developing the narratives at these sites. Yet, there is still a long way to go, and such sites remain contentious.

The creation of monuments and memorials to colonial violence can serve to preserve and promote social memory and national identity. Many sites of colonial violence have received recognition of the site's significance in the creation of Australia's 'nation-building' (Wilson, 2011b, 203). The Monument Australia (2022) website lists 41 public monuments and memorials across Australia that have been erected to commemorate the conflict between First Nations people and invaders from 1788 to the 1940s. The majority of these monuments are located in rural and regional Australia. In many ways, the location of these memorials in remote, or isolated areas facilitates the silence surrounding the violence and death inflicted on Aboriginal communities. Comparing these memorials to sites of convict heritage, for example, it is evident that colonial violence memorials are less frequented, rarely promoted or advertised, and receive minimal national support. Yet, these memorials are present and represent an important aspect of tourism within Australia.

Like other dark tourist sites, visitors to colonial battlefield sites may be motivated by commemoration, entertainment, education and pilgrimage. Research suggests that touring sites of conflict can increase a visitor's empathy with the commemorative space or sacred landscape, challenge us to confront the realities of history, particularly death, suffering and sacrifice, and provide opportunities to learn from, commemorate, mourn and heal (Lemelin et al, 2013). The motivations of tourists will also change depending on the current political and social environment. For example, tourism may decrease if a site is considered 'shameful' but may increase if a site is reimagined as a place of historical significance to be revered and/or celebrated.

Rural sites that explore sites of colonial injustice have several unique factors. Drawing on Hohenhaus's (2013, 142–153) work on commodifying the Rwanda genocide, rural sites can often be less frequently visited, 'isolated' and 'raw' (meaning there is little in the way of tourism infrastructure, which may also mean that it is less colonialised in style), and are often 'first developed as places of remembrance set up and/or looked after by survivors acting as caretakers'. For Hohenhaus (2013, 154), in Rwanda, at least, 'rural memorials ... are genuinely of the very darkest kind of dark tourism. They require a pilgrimage effort of getting there and expose tourists to a much more "difficult" experience'. There is some evidence of this within Australia – some sites of colonial violence are inevitably 'dark' and require travel similar to pilgrimage efforts (for example, the Myall Creek Massacre Memorial); yet other sites, while still difficult to encounter, are much 'lighter' and inauthentic (for example the boab prison tree).

## Myall Creek Massacre and Memorial Site

On 10 June 1838, 28 innocent Wirrayaraay people were massacred by 12 white settlers while camping near Myall Creek station in Northern NSW. Unlike many other massacres and instances of white settler violence, the white settlers were charged with murder and went to trial. Not unsurprisingly, the jury found 11 of the accused not guilty. However, the judge ordered a retrial for seven of the 11 men which resulted in a guilty verdict for all seven men, who were subsequently hanged on 18 December 1838. The enactment of justice to the Myall Creek Massacre led to settlers inflicting 'death by stealth', and the ensuing 'violence, death and dispossession' became 'faint whispers', while a myth of 'benign' colonisation was widely disseminated (Lemelin et al, 2013, 262).

In 1998 a Memorial Committee was established at a community level, which included a descendant of one of the survivors of the massacre, Sue Blacklock, intending to create a memorial at the massacre site. The memorial was opened on 10 June 2000 and was attended by 'descendants of the victims, survivors and perpetrators of the massacre' (Lemelin et al, 2013, 262). The Myall Creek Massacre and Memorial Site was placed on the National Heritage List in 2008 and annual remembrance ceremonies are held at the site. As such, the site attracts both Aboriginal and non-Aboriginal visitors to 'rural regions of Australia' (Lemelin et al, 2013, 267). The memorial features a large rock with a plaque inscribed with:

In memory of the Wirrayaraay People who were murdered on the slopes of this ridge in an unprovoked but premeditated act in the late afternoon on 10 June, 1838.

Erected on 10 June 2000 by a group of Aboriginal and Non-Aboriginal Australians in an act of reconciliation, and in acknowledgement of the truth of our shared history.

We remember them
Ngiyani Winangay Ganunga.

Along the path to the memorial are commemorative plaques, including several 'We Remember Them' messages that tell the story of local invasion – including the repurposing of land for settlement and stock feed, conflicts between local inhabitants and settlers, the killing of hundreds of Aboriginal people which led to some taking refuge at the Myall Creek Station where many would eventually be massacred – as well as plaques telling stories of the individuals that were killed at the site.

These plaques constitute the 'Memorial Walkway', and visitors are 'invited to sit and reflect at intervals as you proceed along the walkway to the memorial rock' (plaque at the Myall Creek Massacre and Memorial site). Further, tourists are asked 'As you walk, please think of the twenty eight men, women and children who died here in brutal circumstances and how you can support a more tolerant future for all Australians' ('Myall Creek – Our Shared History' sign at the Myall Creek Massacre and Memorial site). One of the 'We Remember Them' signs tells visitors that Mounted Police were sent from Sydney to conduct a 'bloody rampage' that resulted in the death of hundreds of Aboriginal people. As such, there is some recognition of the 'urban' interfering with the 'rural', and the location itself highlights the rural dystopia set within the rural idyll or beatific natural Australian bush scenery. This memorial remains remote and private. It lacks tourism infrastructure, meaning that it is likely to receive fewer visitors, particularly tourists seeking entertainment or titillation.

## Incarceration in Western Australia: prison boab trees

The violent conflicts and massacres resulting in illegal land occupation were more pronounced in rural and remote areas of Australia, particularly in the early years of settlement when police were not stationed in remote areas. This meant that colonists were responsible for 'bush justice' and, the 'remoteness from authority, isolation, and fear influenced behaviour, and violence became an accepted way of achieving control over Aboriginal people' (Owen, 2003, 125). Large numbers of Aboriginal people died and a '"conspiracy of silence" characterised frontier relations' (Owen, 2003, 125).

The conflict was exacerbated as Aboriginal people were deprived of traditional food sources and turned to killing stock, such as cattle, as a food source. When police were introduced into these more remote areas, there were insufficient numbers to adequately cover the large geographical areas. Local police were 'forced' to travel significant distances across rural and regional Australia to arrest Aboriginal people who killed stock or damaged telegraph lines. According to Casella and Fredericksen (2004, 117), 'sites of Aboriginal Australian incarceration hold a particularly significant and poignant place in today's Australian society. These sites of confinement often form a radically different physical presence from prisons and establishments in urban centres'. Due to the geographical distances between official police stations, police would chain Aboriginal 'offenders' by the neck and march them to the nearest gaol, often across hundreds of miles over several weeks (Owen, 2003). At times, Aboriginal people, including children, were chained to the natural landscape, including baobab trees in Western Australia (see Figure 2.1, the Derby Boab Prison Tree), during overnight marches (Lowe, 1998; Wickens and Lowe, 2008).

Baobab (also known as boab) trees have attracted considerable attention as being sites of colonial punishment. In India for example, 'executions by

**Figure 2.1:** Derby Boab Prison Tree, Kimberley, Australia

Note: The sign tells visitors that this is a 'site of significance' as a rest point for police and escorted Aboriginal prisoners travelling to Derby as well as its more significant (but less well-known) connection to Aboriginal traditional religious beliefs.

Source: Adwo (nd), 'Boab Prison Tree – Kimberley – Australia'

hangings are carried out from baobab trees' (Wickens and Lowe, 2008, 110). Within Australia, several boab trees have become known as Prison Trees, with reference to colonial police practices of imprisoning Aboriginal Australians *inside* the tree trunks while police transported offenders across large geographical spaces on their way to police stations. Two boab trees in the Kimberley region are analysed here: the Wyndham boab tree and the Derby boab tree. It is important to note that boab trees have a cultural significance for Aboriginal people, providing nourishment as well as sacred burial grounds for deceased members of their community. As such, the 'settler' appropriation of these sites as colonial 'imprisonment' sites further 'darkens' the sites as tourism locations.

## Wyndham Boab Prison Tree

At the eastern end of the Kimberley is the Wyndham prison baobab tree. It has a circumference of 14.7 metres and is estimated to be at least 1,500 years old (Halkett, 2017). There is an opening approximately 60 centimetres (two feet) above the ground which, it has been argued, was fitted with a metal grille to prevent the escape of prisoners (Brooks, 1964). The Wyndham boab, also known as the Hillgrove Lockup was allegedly used as a 'temporary lock-up from the 1900s. It could house as many as 18 prisoners, 30 according to Hill (1934), while its 60 centimetres thick walls and entrance fitted with an iron grille gave little chance for escape' (Wickens and Lowe, 2008, 110). Photographic evidence of the boab tree, taken at some stage between 1917 and 1925 (digitised in 2009) by Mary Elizabeth McCombe (held at the State Library of South Australia) captures graffiti stating, 'Hillgrove Lockup', along with initials carved into the tree (presumably by tourists or locals). This photograph has added credence to the story of the boab being used as a prison tree.

The Hillgrove Lockup was a day's ride from a police station, and as such, the tree allegedly provided additional security from the 'most desperate attempts to escape' (Serventy, 1967, 42). For the boab tree at Wyndham, tourism sites contend:

> Like all countries, Australia holds some dark history. The Wyndham Boab Prison Tree on the beautiful King River Road is one of those examples. Unlike the famous Derby 'Prison Tree', this tree really was used as a prison in Australia's early history as a halfway point and overnight stop between Old Halls Creek and Old Wyndham. Prisoners were captured for things like stealing cattle and were shipped down to Perth, Rottnest and other jails. The police cut the opening into the Boab Tree – the words 'Hillgrove Lockup' were cut into the trunk. (Kimberleyland Waterfront Holiday Park, 2022, para. 1)

Although the Wyndham tree has received recent tourism accolades as the authentic prison tree, in 1967 there was limited awareness of the alleged history of the tree in the town itself. For example, on a fieldtrip to the local area, Vincent Serventy (1967, 42) reported 'it says much for the power of travel literature that most people of Wyndham were quite unaware they had the real prison tree nearby, but knew only of the so-called prison tree at Derby'. Most of the local people in the town told Serventy that he was in the wrong town, re-directing him to Derby. Only one lady and the police were able to direct Serventy to the tree and 'knew' the narrative of the boab being a prison tree.

The reticence of locals to remember or promote the Wyndham boab as a prison tree appears to continue today. There are very few online sites promoting the tree (unlike the Derby tree to be explored next), however, it is located 23 kilometres along the King River Road, which is a rough bush track making the tree hard to access (Bradtke, 2022). The Kimberley Croc Motel (Sherri, 2017) website tells tourists that they will need a four-wheel-drive vehicle (4WD) to reach the tree. Tourists are also warned that the 'Wyndham prison boab tree is of cultural significance to the local Aboriginals. It is fenced off and travellers are asked not to approach it' (Bradtke, 2022, para. 38). As such, tourism to the site is limited, unlike the Derby Boab Prison Tree.

## Derby Boab Prison Tree

The Derby 'Boab Prison Tree', or Kunumudj, is believed to be 1,500 years old (Australia's North West, 2022), approximately 7 kilometres from Derby in the Kimberley region and 2,635 kilometres north of Western Australia's capital city of Perth. The tree has a circumference of over 14 metres, with an oblong split in the bark showing the hollow of the trunk where 'several people could easily stand inside' (Grant and Harman, 2017a, para. 2). Drawing on information from the heritage site itself and the Australian Tourism website in 2002, Casella and Fredericksen (2004, 117) state that the 'Aboriginal people accused of cattle theft were temporarily confined within the tree by local police until they could be transferred to Derby for incarceration'. Other narratives stated that Aboriginal prisoners 'in neck and leg irons, were chained together at the base of the tree', rather than housed inside the tree itself (Bulbeck, 1991, 174).

Today, the Derby boab tree in Western Australia has 'become a major attraction – visited by local and international tourists' (Grant and Harman, 2017a, para. 4). The popularity of the Derby Prison Tree as a tourism site led to a redevelopment of the site in the 1980s as part of the Australian Bicentennial project. The tree is located approximately 500 metres from the Derby Highway, with access along an unsealed roadway. At the end of this road is a car park where visitors are welcomed by an under-cover area that was constructed at the site in 2001 by the Shire of Derby. The shelter includes interpretative panels narrating interactions of early pastoralists and Aboriginal people, the biology of the boab tree and information on Derby and events of the Second World War (Grant and Harman, 2017b).

In the late 1990s, a metal walkway was added to the site to protect the roots of the tree from compaction caused by tourists visiting the site. However, this was removed shortly after complaints were made by local Aboriginal people. In its place, a bush timber fence was constructed around the tree to keep visitors a reasonable distance from it – protecting both the roots from being trodden on and the trunk from graffiti. A sign requests tourists to respect the religious significance of the tree and to remain outside the fence.

The Derby boab tree is recognised as 'a place of great significance to Aboriginal people and the local community in general' (Heritage Council of Western Australia, 2007, 12). The 'selling' of the 'prison tree' was embedded in the Western Australian Register of Heritage Places statement that states that the tree 'represents the harsh treatment [Aboriginal] prisoners often received in the north of Australia in the late 19th and early 20th century' (cited in Grant and Harman, 2017a, para. 4). Online tourism sites advertise the Derby boab tree as having a 'hollow centre and a door cut into its side, the Boab Prison Tree was once used by early police patrols as a staging point for prisoners being walked into Derby' (Western Australia, nd, para. 2).

Cathie Clement also argued that the boab outside Derby was likely used as a temporary cell before the Derby gaol was established in 1887 (cited in Wickens and Lowe, 2008).

Yet, since the 1960s, there have been claims that these narratives are false (Serventy, 1967), and more recently, Grant and Harman's research has suggested that the Derby boab tree (at least) was never used to incarcerate Aboriginal people. Wickens and Lowe have argued that:

> the tree is so close to Derby that the police would have pressed on to the town rather than spend an unnecessary additional night in the open. Since the prisoners were chained together, and had already camped in the scrub for many a night, there appears little reason to shut them up when they reached the hollow tree. (Wickens and Lowe, 2008, 43)

Even if the tree had been used before the Derby gaol was constructed, the lack of Aboriginal support for the narrative of the tree being used to imprison ancestors indicates the tree was never used for this purpose (Kim Akerman cited in Wickens, and Lowe, 2008). Further, an analysis of early newspaper reports, state government reports and official records do not mention police using the insides of the boab trees as prisons, which provides some evidence that these trees have become urban legends. The only 'fact' that documents do support is that Aboriginal people were being detained in the Derby vicinity in the early 1880s (Heritage Council of Western Australia, 2007). Serventy (1967, 42) has also argued that the 'officials in Derby were well aware that theirs was not the true prison tree but no doubt appreciated its tourist value'. As such, tourism to the site (and others) remains, and stories continue to be told of the imprisonment of Aboriginal people in boab trees.

## Creating the myth

The narrative of the Wyndham and Derby boab 'Prison Trees' has a relatively long history within popular culture. In particular, the 'selling' of boab trees as prison trees began in the media in the 1910s. Several stories, across a range of print media, ran articles on the boab trees as 'prison trees', and the number of 'native prisoners' they could accommodate. For example, the Hillgrove Lockup (or the Wyndham tree) could allegedly house 26 'native' prisoners at a time (Sunday Times, 1919; South Western Times, 1923), although, by the 1930s, this number had increased to 30 in the media (Hill, 1934, 43). As previously mentioned, there is also photographic evidence of the words 'Hillgrove Lockup' graffiti on the Wyndham boab sometime between 1917 and 1925, which may mean that this photograph depicts a period *prior* to media accounts of the boab tree being used to imprison Aboriginal people; however, the presence of graffiti does not make something real.

Then, in 1931, the *Queenslander* profiled the Baobab of Western Australia. The very first tree to be featured was the Hillgrove boab, which was given the heading 'A Baobab as lock-up' and readers are told that it was 'used as a prison for natives in the [eighteen-]nineties' (Queenslander, 1931, 29). Almost a decade later, *The Sydney Morning Herald* ran an article titled 'A "Boob" in a boab tree: A queer lock-up' (Ribber, 1940, 9). The slang word 'boob' was used at Queensland's Boggo Road Gaol in 1907 to refer to 'prison' (Boggo Road Gaol Historical Society, nd), and the article utilised this slang consistently:

> Away up in Western Australia's wild and woolly nor-west some distance out of Wyndham there's a boob in a boab tree which is surely the queerest gaol in the world! It was used in the early days for imprisoning natives overnight while on their way to the township for trial and it is known officially as the Hillgrove Lockup. ... Most of its occupants were cattle-spearers and occasionally when large parties comprised the chain gang and there wasn't sufficient accommodation indoors, the natives would be chained to the trunk. But they did not all stay chained. Some years ago a trooper was bringing into Wyndham a party in chains, when at dusk they arrived at the boab-tree. As there wasn't room inside for everybody the trooper chained two of the prisoners to the tree. One of the pair was a magnificent specimen of a man, well over six feet high, with well-shaped arms and legs, and a blacksmith's chest. At daybreak that native was missing, and so was the chain. But an iron bolt to which the chain had been padlocked was left bent back in the form of a hairpin! (Ribber, 1940, 9)

The story draws on several different 'legends' associated with the boab trees: that Aboriginal people were held both inside and were also chained to the outside of boab trees, and that prisoners were eager to escape capture.

A decade later, the *Western Mail* ('Bridle Track', 1950, 14) ran a photograph of the Derby boab with the caption 'Prison Tree: A police trooper when bringing in native malefactors to Derby would frequently use this hollow boab tree as an overnight gaol'. The photograph appeared on the page without an accompanying story but aided in the ongoing narrative that the Derby boab used to house Aboriginal prisoners. The interest in the boab trees continued throughout the 1950s, becoming a popular feature of 'romance and tragedy' fictional pieces for news outlets. For example, Jon Cleary published 'A scream of terror and then murder' in the *Argus* weekender (a Melbourne-based paper). According to the synopsis, 'Justin Bayard, a north-west Australian policeman, captures Emu Foot, an aboriginal murderer who later escapes from his boab tree gaol' (Cleary, 1955, 10). The following decade, the boab featured in the *Australian Women's Weekly* as part of its 'Beautiful Australia' series of articles (Serventy, 1966), combining the rural idyll with the dystopic notion of 'prison trees'; and one writer argued that 'it is now considered less cruel to chain a prisoner to a post or tree instead of confining him inside the trunk of a tree!' (Brooks, 1964, 35). Media stories have enabled a myth to thrive and, in doing so, encouraged continual tourism activity in rural parts of Western Australia.

## The impact of inauthentic tourism

Grant and Harman (2017a) have conducted research on the problematic nature of tourism to these sites and have focused on the inauthentic narrative provided to tourists, the unethical behaviour of visitors desecrating the site and the fact that these trees remain sacred sites to Aboriginal people. For example, despite the erection of fences to keep tourists away from the tree, tourists have shared images of themselves online inside the fence, and often inside the tree itself, showing disregard for the Aboriginal religious site. One site included a photograph of 13 people from a tour group inside the tree (Grant and Harman, 2017a). This photograph of a tour group is particularly unsettling, as it implies that the tour group did little to protect the site or adhere to the directions.

Given the potential income tourists bring to a community, it is important to ask whether it matters if the site is an authentic prison tree. For Grant and Harman (2017a, para. 10), 'the marketing of the tree as a site of "colonial conquest" is sickening. It fails to tell any coherent story of the bloody dispossession of land from Aboriginal people in the Kimberley, a process with ramifications resonating into the present …'. They further argue that the perpetuation of the myth ignores the cultural and spiritual connection Aboriginal people have with the tree. As such, the 'dark' tourist promotion detracts from other authentic touristic potential of the site (Grant and Harman, 2017b), and attracts visitors on false promises, raising ethical questions over the continued promotion of the site as a prisoner tree.

With increasing awareness of the need to confront colonial legacies, and understand the 'truth' of invasion, judicial officers from a variety of Australian jurisdictions went 'bush' to learn about Aboriginal culture (Parke, 2022). As part of this experience, the judges were taken to the Derby 'Prison Tree' where Walmajarri and Nyikina tour guide Edwin Lee Mulligan explained to the judges how the tree 'has become a symbol of the harsh treatment of Aboriginal people in the 1800s and 1900s' (Parke, 2022, para. 19). The narrative presented on the tour was that Aboriginal people were chained to the tree, rather than housed inside, thus, breaking the cycle of misrepresenting the tree while still showcasing it as an important site of colonial brutality. In this way, the participation of local Aboriginal people enables the tree to remain a symbol of colonial violence and settlement without exaggerating the narrative. The chaining of people by the neck, making them march long distances and then being chained to trees at night is surely enough 'darkness' without needing to perpetuate myths of being locked within trees.

The boab trees have, in essence, been mythologised within Australia's cultural heritage. The 'mundane' trees have been converted into the 'extraordinary ... a physical asset ... of spiritual or secular significance' (McKercher and du Cros, 2002, 129). The trees have become a part of Australia's myth of early colonisation. For McKercher and du Cros (2002), mythmaking can help to establish new identities in post-colonial countries to support the political interests of governments. Perhaps, the prison trees were mythologised as a way of reinforcing the notion that Aboriginal Australians were 'offenders' that required being 'locked up'.

## Urban representations

The most common place to locate information about colonial violence in urban areas is within museums and/or art galleries. As popular tourist destinations, museums play a key role in imparting information and providing a cultural repository to tourists. Museums display strategically selected and arranged artefacts and exhibits that are divorced from their 'original context of use and place' (Tunbridge and Ashworth, 1996, 37). Often, this decontextualisation presents a 'particular chosen vision of reality' to visitors that legitimises the actions and ideologies of the dominant political sentiment of the time (Tunbridge and Ashworth, 1996, 37). For many, the potential for museums to 'continue the work of truth and reconciliation' within Australia has fallen short (Manderson, 2008, 13), with the 'dominant ideology thesis' hiding the violence and atrocity of colonisation.

Curatorial politics, often influenced by political agendas, have ensured that narratives have focused on celebrating the 'enduring traditions of an ancient people' (Manderson, 2008, 13), while ignoring, or silencing, the pain and suffering experienced by Australia's First Nations peoples. During the 1980s, the role that museums played in shaping public perceptions of colonial violence and the theft and misappropriation of Aboriginal cultural material came under scrutiny (Harris and Wise, 2018), and has slowly been changing since. Despite fierce opposition, museums have continued to engage in conversations about 'contested histories' and, in particular, colonial violence. Indeed, there has been growing recognition of the misinformation and racial vilification of Aboriginal people within museums, and the Australian Museum has been taking important steps to provide a voice to Aboriginal people.

### Australian Museum

Despite being established in 1827, the Australian Museum's steps towards presenting Aboriginal narratives around colonisation have only recently begun. For example, a First Nations director was only appointed in 2021, and according to the first appointee:

> Historically, Indigenous peoples were not viewed as equal to westerners, and so our bodies, our artefacts and our cultures were collected alongside the animal collections. ... Museums have for a long time controlled our representation, and how we're represented is how we're perceived by the public. (Ms McBride cited in Archibald-Binge, 2021, paras. 4, 7)

As a part of this transformation of the Museum, it was recognised that exhibits needed to address the 'false, constructed history that is pervasively shared in society' and educate the public on the violent dispossession of land and culture suffered by First Nations communities (McBride cited in Australian Museum, 2022a, para. 3). In 2020, a first step towards this was to recognise their role in Australia's colonial history by updating its statement of reflection (which can be found in print on the Ground Level of the Museum) to 'As the first Museum in the nation, established in 1827, the Australian Museum is part of Australia's colonial history and we acknowledge the wrongs done to the First Nations people' (Australian Museum, 2021a, para. 13).

The process of changing the narratives of colonisation within the museum was showcased in a new exhibit, *Unsettled*, which gave control over the 'story' of Australia's history since Captain Cook landed back to the 'original storytellers' (Archibald-Binge, 2021). *Unsettled* recognised that the history of colonisation needs to be addressed truthfully by listening 'to First Nations voices which have been absent from Australia's foundation narratives' (Australian Museum, 2021a, para. 3). The exhibit focused on several thematic sections including 'recognising invasion', 'fighting wars', 'remembering massacres', 'surviving genocide' and 'continued resistance' (Australian Museum, 2022a).

However, there is little reference to geographical differences within the overall exhibit. Three examples within the 'fighting wars' section on the Australian Museum website illustrate this point. While 'The Approach to the Warrego Country Map, c 1845', 'The Sydney wars' and 'The Appin Massacre' are all geographically focused exhibits, there is little reference to their geological significance (two focus more on historically rural/regional elements while the third focuses on urban Sydney) (Australian Museum, 2021b). While a map is provided for Warrego (a geographical area covering a substantial part of inland Queensland and NSW) and the Sydney wars, there is no geographical reference point for the Appin Massacre (inner NSW, now 16 kilometres from the outskirts of Sydney). The Warrego Country map provides several indications of the landscape and potential remoteness of locations. For example, it indicates the Great Dividing

Range, rivers, 'waterless country' and three areas are designated as 'scrub' (Australian Museum, 2021c). In contrast, the Sydney map shows very little information about the landscape or terrain – a few rivers; otherwise, the map is marked solely for battles, attacks, massacres, conflicts and skirmishes (Australian Museum, 2021d). The information provided online for the Appin Massacre focuses on Major General Lachlan Macquarie and Captain James Wallis – in essence the orders 'to clear the land of Aboriginal people and to set examples, through violence and hostage-taking, for the ones openly opposing colonial authority' (Australian Museum, 2021e, para. 6). A few indications of the terrain are made, including extracts from Captain Wallis' diary of attacks: 'I formed link ranks, entered and pushed on through at thick brush towards the precipitous banks of a deep rocky creek, the dogs gave the alarm and the natives fled over the cliffs' (cited in Australian Museum, 2021e, para. 9). However, the rurality of the Massacre is not dealt with in any meaningful manner.

One way for museums to showcase 'rurality' is through the presentation of photographic images. Brendan Beirne's *Dark Days* photographic exhibit is showcased as part of the 'remembering massacres' section of the *Unsettled* exhibit, which features infrared photographs of (mostly) landscapes in a slideshow projected onto a large wall of the exhibit. The photographs 'illustrate the unquiet places where Aboriginal people have been slaughtered' (Australian Museum, 2022b, para. 2). The infrared lens provides a dystopian quality to the beautiful, seemingly tranquil rural landscapes that were once the scenes of massacres of First Nation communities. While there is limited information about the massacres attached to each photo, each massacre is named and dated. The wider 'remembering massacres' section tells viewers that killings and massacres 'were not spontaneous acts of violence on the fringes of "civilisation" – rather these were typically planned and calculated reprisals', instead the order for the 'dispersal' of First Nations communities was a 'codeword throughout frontier Australia for the deliberate and indiscriminate killing of Aboriginal people in systematic and widespread attacks across Australia' (Australian Museum, 2021f, paras. 2–3).

One clear visual geographical demarcation of colonial violence across Australia occurs as part of the 'remembering massacres' exhibit, where a digital map showcasing the colonial frontier massacres between 1788 and 1928 is played on a loop. Red dots appear (in varying sizes) in incremental years since the invasion to illustrate the scale and location of massacres, and at times a sentence describing general definitions and facts about massacres is displayed on the screen to provide brief information (Australian Museum, 2021g). The visual impact of watching a black Australia becoming increasingly filled with red spots (some quite large) is very powerful. While the dots provide some indication of the location, there is no specific information about rural or regional struggles.

A similar map is on display in the 'missions, reserves and stations' section of the *Unsettled* exhibit. Online, the map focuses on NSW, and it is an instant visual reminder of the widespread nature of the forced removal of First Nation peoples and their assimilation (Australian Museum, 2021h). The missions, reserves and camps cover much of the state, with many rural and regional areas featuring. The Museum notes that 'Fringe camps, missions and reserves were often built on unused government-owned or free land near towns; out of sight and out of mind' (Australian Museum, 2021h, para. 5). This provides some insight into the desire of colonising governments to locate missions away from towns and cities; however, there is no discussion about how this relocation away from services impacted communities.

There is slightly more recognition of geographical variations in the 'stolen generations' section. In this section, there is again a map of the eastern side of Australia detailing locations of institutions where Stolen Generations were taken after being removed from their families. Specifically, visitors are asked to 'note how children were moved into large cities and towns far from their homes, making it almost impossible for families to find them' (Australian Museum, 2021i, para. 3). As such, tourists are asked to consider the geographical distance of these institutions, with an intimation of the change of life from 'bush' to 'urban'.

For those tourists unable to visit the exhibit, an online 360-degree interactive tour is accessible via the Australian Museum (2022a) website. The video enables the viewer to walk through the entire exhibit, read the plaques, listen to the audio segments available and even watch the videos or slideshows available at the exhibit. For example, the video 'Living Legaci' (sic) that is played in the exhibit (one wall of the exhibit has the video projected onto it full screen, dwarfing the visitor) features First Nations peoples and beautiful natural landscapes that can be viewed in-full through this interactive tour. Similarly, the tour features audio that oscillates between natural sounds (the ocean, animal sounds, the wind) to traditional singing and even silence. In places, the viewer can click on audio icons which are linked to specific audio stories or information. The narrative of 'land' and 'nature' runs throughout the entire exhibit – yet there are no specific references to rural versus urban. This may be a reflection of the focus of the exhibit on providing a voice to First Nations communities and recognising that there was *no* urban before colonisation.

## Regional representations: McCrossin's Mill Museum

Amanda Nettlebeck researched the difference in curatorial politics between urban and regional museums covering frontier violence in 2011. National museums are expected to create a shared national identity, provide civic lessons and portray history in a positive light. In contrast, regional museums that exhibit frontier stories were free to focus on their local histories in a way that encourages contemplation of different points of view but in a 'non-confrontational' manner (Nettlebeck, 2011, 1124). For Nettlebeck, regional museums may not be as encumbered as national museums on presenting confronting narratives, instead, they may be able to provide 'a local knowledge about frontier warfare that has never gone away' (2011, 1118).

McCrossin's Mill Museum (opened in 1982) located in Uralla, a town in regional NSW, is a relatively small museum that is dedicated to presenting local history exhibits. One exhibit, *A Tribute to the Anaiwan*, is dedicated to the local Aboriginal community. However, contrary to Nettlebeck's hypothesis, this tourism destination in Uralla does not challenge past narratives of colonialism, and instead offers more romantic and traditional narratives of Aboriginal societies pre- and post-colonisation. For example, the script across the entryway suggests that the Anaiwan people had a 'simple existence. ... At one with nature, idyllic even' before colonialism, before the arrival of 'the settlers, the squatters, sheep, cattle, dogs, shepherds, sophistication, grog, exotic diseases, vices, guns'. The focus of this exhibit is on the traditional culture of the Anaiwan people rather than an acknowledgement of the devastation caused by colonialism.

Instead, the impact of colonisation is briefly covered in another exhibit, *She'll be right, mate: Hearts and minds that shaped New England*. Within this exhibit, there are small mentions of attacks on culture and spirituality and state violence dispersed throughout the wider exhibit. For example, different signs tell visitors about the 'shooting of Aborigines', and colonists 'bludgeoning, driving them over cliffs, poisoning. ... The murder of enormous numbers of aborigines [sic] in a "war of extermination"' which was 'an open secret', unreported by the newspapers. While such narratives focus on the local region, there is little exploration of the *rurality* of the region or the longevity of state violence in the region.

## Tensions between cultural tourism and dark tourism

From the 1980s onwards, the Australian government policy documents increasingly discussed '"integrating" Aboriginal peoples into the tourism industry' (Whitford and Ruhanen, 2016, 1081). Part of this has been a recognition that tourists are intrigued by the 'exotic other'. However, the increased emphasis on tourism can also be attributed to the acknowledgement of the need to improve the rights and livelihoods of Aboriginal peoples by preserving ancient cultures and empowering Aboriginal communities (Whitford and Ruhanen, 2016). The recent example of local Aboriginal community members running tours at the Derby boab tree for current judicial officers demonstrates that Aboriginal participation in such sites is growing and enabling Aboriginal narratives to sites of colonial conflict.

None of the sites discussed throughout this chapter (with the exception of the urban museums/art galleries) attract the same level of tourists as convict or penal sites within Australia. This may reflect the more rural, remote locations of colonial violence sites, or it may reflect concerns that the narrative around the violent colonisation of Australia is still too shameful for many tourists to confront (or that such sites lack more formalised and entertainment-based infrastructure). However, it is also likely that it is more difficult to turn a cultural heritage asset based on dissonance and brutality into a highly profitable cultural tourism product, particularly when there has been little social and political support.

Most tourism sites tell a particular story to engage, entertain and inform tourists. Yet the nature and power of the story for cultural and heritage tourism is vital because it imbues the site with 'meaning, bringing it to life and making it relevant' (McKercher and du Cros, 2002, 125). This often makes the site more exciting for the tourist, and having enjoyable experiences often increase visitor satisfaction. The types of stories delivered also 'provide signals about what activities are acceptable or unacceptable at that asset' (McKercher and du Cros, 2002, 125). Stories of massacres, for example, encourage tourists to approach a tourism destination with reverence and empathy. Sites that engage in inauthentic stories that are prevalent within popular culture, such as the Derby Prison Tree, encourage more 'touristy' behaviour such as taking selfies and carving names into the site. Considering the sites explored through this chapter, the Myall Creek Massacre Memorial provides a very different tourist experience to the prison trees, or even to a museum, because the 'needs, wants, and desires' of the intended consumer are different (McKercher and du Cros, 2002, 126).

Museums, in the past, have been required to reinforce national myths, while other destinations have been ignored because they promote a different narrative from that national myth. It is impossible to determine how many visitors throughout history have visited the boab trees because of the narrative of Aboriginal incarceration. However, what is certain is that it is still being used as a 'selling point' to entice new tourists on some online sites and tour companies despite being shown to be inauthentic and offensive to traditional Aboriginal culture and Australia's history of colonial violence and subjectification. In many ways, the use of nature to imprison Aboriginal people provides a sensational and 'triumphant' narrative within a post-colonial context. Using a tree to incarcerate prisoners is more spectacular than a common police or prison building, and it entices people to travel to see such an extraordinary site. The prison tree also provides a more fun, light and entertaining site for tourists.

# Conclusion

While it is clear that colonial history is being 'consumed', and tourists are travelling specifically to experience or understand more of Australia's violent and 'dark' invasion, it is also clear that the narratives being presented are often distorted. As Casella and Frederickson (2004, 102) have argued, concepts of heritage, culture and national identity are 'rooted in shared memories of a past that are both artificial and romanticised'. The romanticisation of the building of Australia's past, and the colonial myths and 'fantasies' of colonised landscapes and peoples 'continue to play a powerful role in shaping the current gaze of tourists today' (Tucker and Akama, 2009, 11). Indeed, as Bulbeck (1991, 170) argues 'most monuments avoid the sore spot of race relations, the moment of contact, by confining Aboriginal history to prehistory'. Urban landscapes feature monuments of cultural Aboriginal icons, such as canoes or corroboree trees. These monuments serve an important role in recognising the traditional owners of the land, yet they also repress the history of conflict and violence (Bulbeck, 1991).

While this process is changing, there is clearly much more progress needed. For McKercher and du Cros (2002, 115), 'the conversion of a cultural asset into a cultural tourism product necessitates the transformation of that asset into something that can be consumed by the tourist. This process is normally achieved through some level of modification, commodification, and standardization of the asset'. While this process can threaten cultural facilities, it has been argued that the benefits of tourism (including the commercial industry and the tourists themselves – for example, learning or creating empathy) necessitate this transformation. Or simply put, without the transformation would there be tourism?

Taking the example of the boab trees, if they had not been promoted as natural prisons, would tourism to these areas be anywhere near current levels? The new narrative of the trees as sites of physical imprisonment to the *outside* of the trees still facilitates the dark tourism model and may detract from understanding the original cultural significance of the boab trees. Yet, if Aboriginal communities *want* to educate tourists on the use of boabs to hold prisoners overnight and use these trees to symbolise the pain and violence inflicted on their ancestors by settlers, then this should be a way of empowering local Aboriginal people and allow them to transform the cultural asset into cultural tourism.

Transformations of narratives dealing with colonial violence are occurring across Australia. The narrative at the boab trees is more extensive and inclusive of Aboriginal people. Many rural and regional areas have their own memorials to colonial massacres that have been created or shaped by local Aboriginal communities, and urban, and to a lesser extent, regional museums are shifting their narrative. As Manderson (2008, 12) argues, museums and perpetual commemorative sites provide 'a pathway to healing for Indigenous Australians'. Rural and regional museums are also participating in the important work of repatriation of Aboriginal artefacts. For example, the Burke Museum in Beechworth, north-east Victoria has returned numerous objects to the Kaurna people that it has held since 1878 (Smyrk, 2022).

Dark tourism sites and exhibits that deal with colonial violence against Aboriginal Australians provide an opportunity to educate tourists on the experiences and perspectives of First Nations peoples and their descendants. None of the rural sites discussed here provided overt tourism infrastructure – at most there were interpretive signs and, at the Derby boab tree, a guide for a very specific group of people. Some tour groups take tourists to the Derby boab tree, however, none of these sites offers tourists memorabilia (except, of course, in the form of photographs and memories). Museums offer the most tourism infrastructure, and places like the Australian Museum are taking active steps to ensure that colonial violence is presented from an Aboriginal perspective, and while the narrative is undeniably (and naturally) 'dark', it is dealt with in a respectful and empathetic manner.

The coverage of rurality as a theme, however, is almost absent. Tourists travelling to on-site rural memorials and the boab trees will inevitably be exposed to the landscape and remoteness of the site. They are more likely to have a deeper understanding of what it meant for Aboriginal people to be chained at the neck and marched hundreds of miles throughout the year in all different types of weather. However, this distinction is lost within the urban setting of a museum. Thus, visiting colonial Aboriginal 'dark conflict sites' in situ provides an important historical and cultural experience for tourists.

While colonial violence sites are not key tourism destinations, either for national or international tourists, their existence effectively provides a physical history of colonialism and can help in the reconciliation process within many local communities. Colonial violence sites are likely to remain uncommercialised because the process of converting an Aboriginal cultural site into an economically viable tourism site will likely increase perceptions of unethical and inauthentic tourism. Where such sites are developed, Aboriginal communities within the local community should be at the forefront of the discussions, such as in the Myall Creek Massacre site. Not only does this enhance the sense of authenticity associated with a site, but it also provides an avenue for further healing and reconciliation for both the community and tourists.

The Australian Museum provides a useful model for integrating Aboriginal voices into tourism exhibits. While the geographical rural element is often lacking in such exhibits, perhaps the first step in encouraging tourism to (and therefore education and understanding of) colonial violence sites is to increase the presence of such exhibits in *urban* environments. Employing local (and differing) Aboriginal community elders and members to provide descendants' oral histories and understandings of colonial violence will move beyond a 'tokenistic', Europeanised representation of the colonisation of Australia (Foster, 2020). Of course, Aboriginal communities must want to engage in this process – this should not be enforced – and indeed, if the Aboriginal community do not want to engage in the process, then colonial violence should be omitted until such a time as an inclusive narrative can be provided to tourists.

# 3

# Convict Tourism

Australia's cultural heritage is intricately linked to its convict past. Australia's colonial history has two dominant romanticised archetypes: the rural pioneer and the law-breaker (Carroll, 1992; Casella and Fredericksen, 2004; Jones, 2016). Convicts are often seen as embodying both these roles, with all early convicts essentially transported to 'rural Australia'. The term 'convict' 'has assumed an iconic status in the national gaze unparalleled anywhere else in the world' (Casella and Fredericksen, 2004, 104), and as such, makes it a desirable tourism drawcard.

More than 162,000 convicts were transported to Australia from England, Ireland, Scotland and Wales over 80 years between 1788 and 1868. Convicts were transported to two major convict colonies: Sydney, NSW, from 1788 to 1840, and Van Diemen's Land (later renamed Tasmania in 1856) from 1803 to 1853 (Maxwell-Stewart and Oxley, 2020). Later, convicts were also transported to Swan River (Western Australia) from 1850 to 1868 (and this has since become an important convict heritage tourism destination, partly due to the imprisonment of Fenian political prisoners sent there during this period). In addition to these transportation sites, several satellite colonies or penal settlements (also known as 'secondary punishment' sites for colonial offending that was punished by 'transportation') were established at Port Phillip (now Melbourne), Moreton Bay (now Brisbane), Norfolk Island, Newcastle, Port Macquarie, Macquarie Harbour, Maria Island and Port Arthur (Maxwell-Stewart and Oxley, 2020). Several of these sites, such as Port Arthur, have become well-known tourism destinations because of their role as 'secondary punishment' sites.

While the convict system in Australia was not predominantly characterised by incarceration and institutionalisation (Jones, 2016), it appears as though those buildings that did incarcerate convicts remain popular tourist destinations that resonate with the popular fascination with an *imagined convict past* (Casella and Fredericksen, 2004, 105). The reality of the convict experience is that they (and free settlers) were transported to provide human labour resources for the creation of a colony, which inevitably meant that the work they performed was often physically taxing and involuntary, and often accompanied by food and clothing shortages (Wise and McLean, 2021). Transportation meant exile, and 'it was a fierce punishment that ejected men, women and children from their homelands into distant and unknown territories' (Bogle, 2008, 23).

The experiences of convicts depended on a range of factors, including the location they were sent to, their original crime, how willing they were to work and their behaviour. Across these institutions, 'normal' punishment could include physical reprimands (flogging) or hard labour in chain gangs. However, in some locations, convicts could also experience solitary confinement among other cruelties. Despite this, research into the convict era confirms that only a minority of convicts suffered brutally, and while convicts experienced physical and emotional pain, it 'was limited in time and did not dominate their lives' (Smith, 2009, 313).

With the end of transportation in the mid-nineteenth century there were clear political and social efforts to distance Australia's future from the 'convict stain' of its past. Many convict establishments were dismantled or repurposed with the intent of 'forgetting the past', even to the extent of renaming entire locations in an effort to dissuade unwanted tourists. Yet despite these efforts, some sites became instant, and much sought-after, tourist destinations.

In the 1970s, it became obvious that tourism was not being deterred, and indeed could become a profitable tourism industry. At the same time, there was a wider political and social shift which saw a change in discourse around Australia's 'unsavoury historical incident' – that of the 'convict stain' (Jones, 2016, 26). That discourse transformed into a new narrative of convicts as Australia's 'reluctant pioneers' (Barnard, 2010, 7), which needed to be celebrated. As a result of this rebranding, many convict sites became more commercialised and an increase in tourism infrastructure was invested to preserve Australia's culture heritage. As such, in the Australian context, 'penal museums ... are particularly popular cultural institutions, not only among tourists but locals with whom the museums resonate as part of a mythologized convict past' (Walby and Piché, 2011, 454). Further:

Penal colonies form an important historic cornerstone in colonial Australia, and therefore are considered part of Australia's widely debated uncomfortable heritage. ... In the last few decades, many of those decommissioned Australian gaols listed as heritage buildings have undergone adaptive reuse. They have been transformed from uncomfortable and shameful memories to community spaces or tourist attractions. Most of these gaols were adapted to museums that celebrate the dark history of the site, while in a few cases, preserved gaols were integrated with mixed-use and residential developments, reused as boutique hotels, event venues, theatres, or art schools. (Shehata et al, 2018, 1)

Convicts are now celebrated within Australia's history. According to the 'WHO DO YOU THINK YOU ARE?' poster at the Hyde Park Barracks, the popular shift to interpret convict ancestry as a badge of honour, rather than a 'stain', occurred during the 1970s (see also Welch, 2012 for a more detailed discussion). Even so, most sites still engage the visitor with uncomfortable and 'dark' exhibits and information surrounding this history. Despite this, or perhaps because of this 'darkness', such sites offer a constant supply of visitors each year.

While more than 3,000 convict sites remain across Australia, only 11 make up the *Australian Convict Sites* World Heritage list, and each site tells:

> a story of exile from one side of the world to the other and how a new nation was formed from hardship, inequality and adversity. Together the sites represent the global phenomenon of convictism – the forced migration of convicts to penal colonies in the 18th and 19th centuries – and global developments in punishment of crime in modern times. The *Australian Convict Sites* are the preeminent examples of our rich convict history. ... This is unique in the world today. (Department of Agriculture, Water and the Environment, 2021, paras. 1–3)

Of these 11 sites, five are located in rural and regional areas (Brickendon and Woolmers Estates, Coal Mines Historic Site, Darlington Probation Station, Kingston and Arthur's Vale Historic Area (Norfolk Island), Port Arthur Historic Site), and four of these are in Tasmania (with an additional 'urban' convict site within Hobart).

Using Tasmania as a case study, this chapter will highlight not only how small communities can be burdened with unwanted historical heritage but also how this disdain can shift and such sites can be celebrated and converted into some of Australia's most well-known and visited dark tourist sites which actively incorporate the landscape and 'rural idyll'. Tasmania boasts the record for its number of world heritage recognised convict sites within Australia (with five listings compared to four in NSW, one in Western Australia and one in Norfolk Island) and has invested large sums of money into the protection and reconstruction of its sites. A large part of this reconstruction has involved creating extensive tourism infrastructure that provides crucial economic support for local, rural and regional communities. In addition, because of their remote location, many of these sites are very difficult (and expensive) to access, yet people are willing to travel significant distances to immerse themselves in this part of Australia's cultural heritage.

# Port Arthur Historic Site

Within Australia, the Port Arthur Historic Site is one of the oldest (Design 5 Architects, 2003) and best-preserved convict sites. Further, Wilson (2011b, 205) argues that it is 'one of the most popular dark tourism sites' and also 'the most visited by schoolchildren', making it an important site to visit and research. In 2017–18 there were 368,862 'day visitors' to the Port Arthur Historic Site (PAHS) (an increase of 9.6 per cent from 2016 to 2017) (PAHSMA, 2018). Before COVID, approximately 54 per cent of tourists to the site were Australia-based independent travellers (constituting 14 per cent from NSW and 15 per cent from Victoria); 20 per cent were from cruises (including international and national travellers); 18–19 per cent were independent international tourists; and 9 per cent were independent Tasmanian tourists (Parliament of Tasmania, 2020). As a result of the COVID pandemic, 2020 saw a decrease of 24 per cent in numbers to the site and the closure of the site for some time (Parliament of Tasmania, 2020).

The site won the 'Australian Traveller People's Choice Award' for Best Historical Site in Australia and was also recognised with a Silver award for Cultural Tourism in the Australian Tourism Awards in 2017–18. By 2019, the site had won gold at the Australian Tourism Awards under the Major Tourist Attractions category and two Gold Awards at the 2018 Qantas Australian Tourism Awards for Major Tourist Attractions and Cultural Tourism. The Port Arthur Historic Site has 'local, state, national and international heritage significance' (PAHSMA, 2009, 79).

Set on the southern tip of the Tasmanian peninsula, Port Arthur was first established in the 1830s by white settlers as a penal settlement for British convicts. Point Puer, the juvenile reformatory located at Port Arthur, has been described as 'Australia's first rural reformatory' (Jackman, 2001, 6). The penal site is located in what was commonly described as a harsh and hostile environment, and the prison was known to have instituted brutal punishment on exiled prisoners, with early museum narratives going so far as to describe the place as 'Hell on Earth' (Strange, 2000a, 4).

## Port Arthur as a convict settlement

On 20 September 1830, the *Derwent* brig arrived at Port Arthur in Van Diemen's Land with 15 soldiers and 30 convicts to establish a secondary punishment timber station (Pridmore, 2009). England selected the Tasman Peninsula for its penal colony 'for its remoteness and isolation. The peninsula is connected to mainland Tasmania by a slender isthmus known as Eaglehawk Neck, which is less than 30 meters (33 yards) wide. Aside from this narrow land link, the Tasman Peninsula is surrounded entirely by water' (Mason et al, 2003, 4–5). While Port Arthur certainly had physical 'prisons', it also relied on the natural, rural, landscape to control and contain convicts. As historian Maxwell-Stewart notes:

> In the nineteenth century all official communication with Port Arthur was by sea. The only land connection, the route that terrestrial absconders would have to take, passed through two narrow necks. Lines of dogs, whose bark gave away the presence of all would be escapees, were used to seal these. The Tasman Peninsula was thus a natural prison. (Maxwell-Stewart, 2013, 24)

Port Arthur had been designed as a penal station for reoffending or 'hardened' repeat offender convicts (Maxwell-Stewart and Hood, 2010, 5) who were consequently required to serve a more severe sentence. Despite the transportation of prisoners ceasing in May 1853, the site remained a prison until 1877. It is estimated that Port Arthur housed 10,000 convicts, and 12,600 sentences were served (some convicts served multiple sentences) while it was open (Pridmore, 2009).

Port Arthur remains an important heritage site for Australia, not only as a convict establishment, but also because it offers a unique physical example of the ideologies of punishment that were occurring in the British Empire during the late-eighteenth and early-nineteenth centuries. The reform movement, focusing on improving criminals and thus reducing criminality within society moved away from flogging as a form of punishment and instead introduced notions of labour, strict discipline, theological instruction, separation and silent treatment where the prisoner could reflect upon their behaviour.

In line with this ideology, Port Arthur adapted solitary confinement as a form of punishment from at least 1837 (Design 5 Architects, 2003). However, as these cells still allowed communication between prisoners, Dr John Hampton, Comptroller of Convicts, announced in 1846 that a prison should be constructed in Van Dieman's Land based on the Pentonville Prison in London. Construction subsequently began on the Separate Prison (also known as the Model Prison) at Port Arthur in 1847 and began to be used in 1849 before its completion in 1852 (although there were further additions in the years to come). The Separate Prison at Port Arthur was designed for 'incorrigibles' – 'the very worst class of reconvicted men' and was 'designed to produce docility in those considered dangerous' (Design 5 Architects, 2003, 17).

Despite the 'good' intentions of the reforms, the Separate Prison was the most brutal and inhumane area within the Port Arthur penal settlement and has been referred to as the 'terror' of the settlement and the 'most dreadful of penal institutions in Australasia' (Beatties Studio, 1990, 21). Not surprisingly, the Separate Prison is one of the main 'attractions' for tourists visiting the site.

## Early tourism

The tourism activities described in historical accounts seem to have a different focus from modern-day tourism to Port Arthur. Historical tourists to the site, particularly those travelling for the Boxing Day picnics, reflected on the whole of the experience (including the lengthy boat crossing). From these reflections, it is clear that *entertainment* was the sole focus of the tourists' desire to visit the site (rather than trips for memorial purposes). However, a common feature of tourism to the area remains the geographical isolation of the site and the time and resources required to visit this rural location.

Visitors travelled to the Port Arthur convict settlement while it was a functioning prison to witness first-hand the true 'convict experience'. For example, David Burns was allowed to visit the site just 12 years after construction, in 1842. His visit entailed a full tour of the buildings and grounds, including access to the convicts themselves. In another account, the authors of the 1869 Melbourne publication *Guide for Excursionists from the Mainland to Tasmania* outlined their trip to Port Arthur (who were also allowed to view the convicts as they worked and slept) and recommended it as a site for wealthy tourists if they could obtain the necessary permits (Jones, 2016).

The closure of the site in October 1877 provided tourism opportunities, 'with people of all classes, from the respectable men prompted by curiosity to see the detainees, to the association of the prisoners themselves, anxious to get a look, or if possible effect an interchange of passing words' watching the relocation of the remaining prisoners board boats to Hobart (Tribune, 1877, 2). Following the closure of the site, Port Arthur became a prominent tourism site, with tickets for the day excursions from Hobart selling out quickly, with the demand for travel exceeding the boats' capacity. For example, on Boxing Day in 1877, 800 'excursionists and pleasure seekers' were taken to the penal site by a steamboat called the *Southern Cross* (Tribune, 1877, 3). The tour to Port Arthur was so popular that it was estimated that another 200 tourists were left behind at the docks (this trend lasted many years).

Many of these historical accounts conveyed the geographical remoteness of the site and the preparations required to visit Port Arthur. Indeed, for many of the early tourists to the site after it was decommissioned, the *journey* itself to Port Arthur was just as important to them as the site itself, in terms of whether they considered the tour a 'success' or 'pleasurable'. Tourists recounted the weather, the conditions of the sea, the company and how tourists were treated by the crew of the relevant steamer (Mercury, 1881). Even as early as 1890, tourists from across Australia were travelling to Port Arthur (Church of England, 1890), and the American writer, Mark Twain, visited Port Arthur in December 1895 (PAHSMA, 2009). It was estimated that 5,000 tourists visited the town in 1912 (Context Pty Ltd et al, 2002).

Day trips from Hobart were relatively inexpensive – adults were charged five shillings and children cost two shillings six pence in 1883 (Young, 1996), while some travellers spent at least one night in Port Arthur and paid more for this experience (12 shillings return trip on the *Pinafore* in 1885). It is important to remember here that there was no tourist accommodation available at the site until 1885 when the Commandant's Residence became the Carnarvon Hotel (Context Pty Ltd et al, 2002). According to Young (1996), day tourists were less respectful of the convict site and were frequently accused of 'merrily dancing', locking themselves in cells and collecting souvenirs in the form of moveable objects.

In addition to the cost of the travel, Port Arthur remained a relatively hard place to visit, and as previously mentioned, the demand for such tourist activity exceeded the capacity of tourism providers. According to one report, only three to four trips were arranged by sea to Port Arthur each year (Argo, 1912) and these usually occurred in the summer months (Context Pty Ltd et al, 2002). Steamers from Hobart offered the most direct route, with a travel time of around three hours each way. From tourist accounts, the length of the trip to and from Port Arthur took up the majority of the 'tour', leaving usually only a few hours at the site itself. The interest in Port Arthur certainly never seemed to wane, even in 1900, steamers were leaving behind numerous passengers due to limited passenger capacity, with one article estimating that on one trip alone £10 of income from potential tourists was refused (Mercury, 1900).

There were also private tours of the site, and throughout the 1920s, the site was visited by travellers across the land. In 1924, the *Age* reported that a Victorian football party took eight motor vehicles to the site and hired a guide (Age, 1924). By the mid-1920s it was becoming more common for tourists to be travelling significant distances to visit the site. For example, one tour group consisted of tourists from Victoria, NSW, Queensland, America, Italy and Belgium (Grapho, 1925). Towards the end of the 1920s, Port Arthur was so popular that it attracted 'big crowds all the year round' and kept 'a small army of motor drivers and others in employment' (Herald, 1928, 10).

Once at the site, visitors maintained their expectations of entertainment. While modern visits to Port Arthur are focused on memorial understandings, with a side of entertainment, the historical focus hinged on providing a memorable and, in particular, *fun* experience. For example, in 1880, just three years after the closure of the site, tourists recounted with delight how the Artillery Band took position in one of the rooms of the Model (Separate) Prison and played 'such sweet music that many young folk could not resist the desire to dance, a pastime which was carried out with zest' (Mercury, 1880, 3). The same prison was often a source of delight for tourists who could be 'locked up' (sometimes unintentionally) in one of the dark cells:

> model prison ... [was] the centre of attraction, in which place some amusement was caused by two young ladies going into the cells, and the doors being closed, they were made prisoners until some kind friend went in search of ... the keys, and thus released them. (Mercury, 1880, 2)

Part of the 'pleasure' experienced by tourists was the guides employed to show tourists around the site. On some tours, the guides were former wardens, and at other times, former convicts. The desire for a guide who intimately knew the site was frequently commented upon in these tourist accounts, ' "Alf", the Government guide, who shepherded us through those sinister landmarks of the settlement, could keep you interested for days on end. He is saturated with convict lore and with memories of the past' (Rolph, 1935, 4).

In another example, one tourist wrote 'the great ambition of the average tourist visiting Port Arthur, is to discover one of the old hands, or legs, as they were commonly called, i.e., convicts' (G.T.H, 1910, 2). Convicts that could tell gruesome stories were frequently sought after, and by 1892, the owner of the Model Prison, Reverend Woolnough, began officially charging admittance to this building and employed an ex-convict from the site to provide information on 'ancient horrors of the place' through guided tours (Young, 1996, 70). In addition to live music and 'old hand' guides, tours also offered tourists picnic opportunities.

While not organised in any sense, some tourists took the liberty of removing 'relics' from the site:

> What had we for our pains? Well, the relic-hunters amongst us were not wholly empty-handed. ... A piece of plaster from the wall of a punishment cell, a rusty vessel picked up in the [unclear] of the kitchen, from which some old-time convict may have [unclear] 'skilly', a fragment of iron from the peephole in a cell door, a splinter of wood from the worm-eaten pulpit in the prison chapel. The depredators entertained no conscientious scruples as to their right to these mementoes. ... Too many thousands of shillings would be required to put the model prison into anything like trim condition ... besides even were such a measure contemplated, would not the removal of a little of the rubbish encumbering the threshold rather facilitate then retard operations. Therefore the conscience of the pillagers [sic] was at rest. (Argus, 1890, 9)

In addition to the pillaging of the site, these same excursionists participated in a 'mock auction' conducted from the pulpit of the church in the Model

Prison. Consequently, it appears, particularly in the early years of tourism after Port Arthur closed, that tourists expected their trip to consist of merriment, pillaging and generally having fun at the site.

## Carnarvon – the distancing

Although Port Arthur ceased to be a penal settlement in October 1877, the reality is that the site was slowly abandoned from 1871 when the British government transferred control of Port Arthur to Tasmania. According to Jones (2016), the convict settlement was a source of embarrassment for many respectable Tasmanians, and it seems little upkeep was maintained from this point, with many buildings falling into disrepair by 1876. Despite the clear tourist interest in the site, the local community and government wanted to distance the place and people from its historic convict origins. Just two days after the first 'Boxing Day' excursion to Port Arthur, which saw tourists causing destruction and vandalism, the first lot of land at Port Arthur was auctioned to private buyers (Brand, 2003; Jones, 2016). While there was limited interest for this first auction, further land was sold in 1884 and 1889, with the site being renamed Carnarvon in the 1880s:

> A new township named Carnarvon was superimposed on the remains of the former penal settlement. Much land within the former penal settlement was subdivided for farms and orchards during this period, which created new settlements across the Tasman Peninsula. Small rural settlements grew out of the former probation stations, and Carnarvon became a crossroads town and the centre of community life for the Tasman Peninsula. (Context Pty Ltd et al, 2002, 31)

In 1872, Attorney General Dobson led a campaign to 'destroy all convict buildings and relics' and when the bushfires occurred, they were seen by many as a welcome 'purifier' (Daniels, 1983, 3). As the township of Carnarvon grew and expanded, the penal buildings associated with the convict site were adapted, rebuilt and removed. By the early 1900s, the asylum had become the town hall, many new buildings were created and by the 1920s, 'Carnarvon had the appearance of a neat rural village' (Context Pty Ltd et al, 2002, 31).

Walter Pridmore (2009) argues that the location was renamed to dissociate the township from its penal past, and some (though definitely not the majority) of the newspaper articles supported the move to rename the site:

> Port Arthur is not to be found as a township, as its name, with all its disagreeable associations, belongs to the past, and in its place we see Carnarvon, one of the most picturesque little villages possible. (Illustrated Sydney News, 1891, 18)

In 1913, a news editorial recommended that the settlement ruins:

> be razed and cleared away entirely and its site used for some edifice
> of more aesthetic appearance, and pleasanter associations. ... We
> need memorials and reminders that are cheerful and inspiring, not
> depressing, humiliating, saddening. ... Men rise on stepping-stones
> of their dead selves, and need not have those ugly corpses hung round
> their necks or sitting at their tables. (Hobart Mercury, 1913, cited in
> Frow, 1999, 2)

In 1893, a volunteer group, the Tasmanian Tourist Association, formed to
promote Tasmania as a tourist destination, including Carnarvon, though
the focus was on the scenery and rarely on the convict heritage of the site
(Mason et al, 2003).

However, some vocal tourists writing within the papers expressed
their distress and saw the commercial benefit of retaining the site as a
tourism drawcard:

> unmitigated disgust ... at the huckaterug meanness and the callous
> Philistiniam of the Tasmanian Government in selling the Model Prison
> at Port Arthur, and selling it for £850! ... That the convict buildings
> at Port Arthur are exceedingly interesting is proved by the increasing
> number of travellers who undertake the journey from Hobart to
> inspect them, even with the existing difficulties of conveyance. ... If
> money was the object of the Tasmanian Government why did they
> not retain the prison and charge 1s a head to all the summer visitors
> who wish to see the curious interior? They would have then acquired
> a much larger revenue than they can obtain from the sale of the paltry
> sum for which they sold the building. ... As the numbers of visitors
> to Port Arthur annually grows so will the feeling of indignation grow
> against the vandalism of the Tasmanian Government. (Black, 1890, 4)

Indeed, it seemed the financial benefit of the prison was recognised
early within the community, and the township quickly capitalised on its
convict history when it became apparent that tourism offered a lucrative
economy. According to Mason et al (2003, 7), the 'first concerted effort to
benefit financially from the site's tourist potential came in 1881' when the
Whitehouse brothers launched a biweekly steamer service between Hobart
and Norfolk Bay, which transported visitors to Carnarvon. Similarly, when
the government again placed the major buildings of Port Arthur for auction
with the condition that the structures be demolished after purchase, many
local community members protested this destruction on the basis of losing
tourists (Jones, 2016).

The making (and then remaking) of Marcus Clarke's book *The Term of His Natural Life* into a film, first in 1908, and then again in 1927, highlights the ongoing dilemma within the community over the perceived benefits of tourism to the local areas and the desire to shy away from the dark and 'disgraceful' past. For example, while the 1908 movie (not filmed on site) saw a boost in tourism to Port Arthur, there was still trepidation over the 1927 remake being filmed on site, causing controversy and objections that it was 'a bad advertisement for Tasmania', and newspapers suggesting that it would spread 'an outrageous lie' internationally (Smith, 2009, 310). Despite those concerns, the film was a success and helped to drive more tourism to Port Arthur.

As tourists continued to travel to the site to 'see first-hand the "horrors" of a penal station' (PAHS, 2019), the Carnarvon community quickly established a 'private museum, guided tours (often offered by "old lags"), the sale of souvenirs and the provision of accommodation catered to tourists' interests and created a financial base for the community' (PAHSMA, 2009, 28). In 1927, the site was again renamed Port Arthur. A number of factors including the introduction of motor vehicles, improved roads and access, and a renewed focus on entertainment and recreation saw tourism to Port Arthur continue to grow (Context Pty Ltd et al, 2002).

To provide further context to the rurality of this location and the importance of tourism for Carnarvon, in 1884, the local population of the *whole* of the Tasman Peninsula was approximately 600, and 'while relations between tourists and locals were at time strained, locals recognized the importance of tourism to their livelihood, even indirectly through the sale of produce and goods' (Jones, 2016, 43). Tourism meant that roads were repaired and upgraded, motels were established and liquor licences gained, all of which benefited local residents.

## *Tourism forces a reconnection to the convict past*

Tourism at Port Arthur clearly has a troubled history. Tourists visited the site to 'watch' convicts, and then they continued to participate in 'merry-making' and 'souvenir collecting' of the site once it was closed. The community (and government) initially had clear designs to distance itself from the convict 'stain' of the era and start fresh under the name Carnarvon. Yet, tourism remained dominant, forcing social and political ideologies to adapt, reinstate and protect what remained of the convict heritage. It became very clear that a more organised approach to tourism in the region was needed.

In 1913, the Tasmanian Tourist Association formally requested government support to manage the ruins. However, as already discussed, the site had suffered damage and decay even before the site was closed, which was then followed by destruction and vandalism by tourists, then two fires in 1895 and 1897, which destroyed many of the buildings in Carnarvon, and also severely damaged the Penitentiary, Separate Prison and hospital from the original convict site. As such, considerable work was required.

According to Jackman (2001), Port Arthur underwent three major conservation paradigm shifts following the period of fire destruction in the 1890s. The first, Jackman (2001) claims, occurred during the period 1913–40 when the Convict Church, Guard Tower and Government Cottage were stabilised. At this time, tourism remained the town's most important source of income, and by the 1920s and 1930s, the town held three hotels and two museums (as well as guided tours) (PAHS, 2019). As a result of the tourist attraction, a Scenery Preservation Board (SPB) was created in 1916 to manage the penal site, and visitors to the site began to be regulated in 1926. The SPB created five reserves within Port Arthur: the church, the penitentiary, the Model Prison, Point Puer, and Dead Island, which became Australia's first gazetted historical sites (Mason et al, 2003). During this period, the Soldiers' Memorial Avenue was established in 1919 to enshrine the memory of local men who were killed in the First World War (PAHSMA, 2009).

During the second paradigm shift, the Tasmanian State government purchased the town of Port Arthur for £21,000 in 1946. Tourism continued to grow after the Second World War, particularly as access to the site remained free with small fees for guides (Mason et al, 2003). By 1970, the government owned much of the site and in 1971 the site came under the management of the Tasmanian National Parks and Wildlife Service. The focus of reconstruction and preservation during the mid-1970s included the Penitentiary, Separate Prison and Asylum, reflecting the same period of Australia's shift in accepting and celebrating convict heritage. In 1979 $9 million (unless otherwise stated, all currency refers to AUD) was injected into the site to conserve, develop and 'put the site into order and prepare it for display' (Staples, 1995, 38).

Yet, criticisms continued that the site failed to explore 'its role as a penal station within the context of the nineteenth-century debate over transportation and the related rise of the prison' and that evidence of convicts, such as cells, was glaringly missing (Maxwell-Stewart, 2013, 25). Daniels provides the following caption from the museum at Port Arthur to highlight the message about convicts during this period:

No aspect of Port Arthur's history has been more distorted than the punishment and treatment of prisoners. Most of the horrifying stories that abound have no basis in fact and the cases of brutality that did occur seem to have been committed mainly by convict trusties against their fellow prisoners. (Daniels, 1983, 6)

From this, the convicts were portrayed as the 'problem', while the institution was depicted as humane, demonstrating the power of 'narrative' within tourist sites and the influence of political and social ideologies. Daniels (1983, 6) also quotes a tour guide from this period telling tourists that convicts were '*better off*' at Port Arthur than they would have been 'back home' because they were fed and housed (freedom and family connections were not mentioned).

The third stage, spanning the mid-1980s to early 2000s, was principally concerned with civil-officer accommodation, precinctualisation and commercialisation. During this period, the *Port Arthur Historic Site Management Authority* Act was passed in 1987, transferring authority of the site to the Port Arthur Historic Site Management Authority (PAHSMA). Throughout this period, Mason et al (2003, 26) refer to three 'pendulum swings' that management took in 1985, 1996 and 2000, 'in order to balance conservation and the access/tourism activities required to operate the site'. Piper (2016) argues that this period saw a failure by the federal and state governments to provide effective long-term conservation to Port Arthur, particularly in the context of climate change and rising sea levels. The consequence of this was little to no conservation work on the site, despite increasing profits because of tourism. Instead, the funding supported growing tourism infrastructure. The 1996 massacre at the site (see Chapter 6 for further details) led to an injection of funds – but not for the preservation of the site. Rather, the funds were to provide further tourism infrastructure in the form of the visitor centre and car park, highlighting the focus on accommodating tourism rather than the preservation of cultural heritage (Piper, 2016).

## Organised tourism – Port Arthur as a 'dark dungeon' and a 'dark fun factory'

For Jackman (2001, 11), these three distinct paradigm shifts have inevitably affected the site conservation and interpretation, with the first conservation reflecting 'creeds of moral-force cum-civic-duty' mantra, the second a 'government collectivisation and institutional power' and finally, a 'personal and domestic empowerment and "user-pays" economic rationalism of the 1980s and 1990s'. In particular, Jackman is highly critical of the PAHS of the late 1990s:

> The cumulative effect is its present incarnation as a pay-to-enter convict theme park, featuring romantic ruins in a garden landscape that supports a gothic-horror cultural tourism industry. Commercialisation and marketing of products such as ghost tours, sea planes, electric people movers, costumed role plays, à-al-carte [sic] dining, pyrotechnic festivals and other entertaining diversions obstruct a more critically

meaningful engagement with the place, and totally disconnect it from any contemporary social debate on crime, punishment and rehabilitation, and the maintenance of hierarchies of privilege in society. (Jackman, 2001, 11)

Piper's (2016) analysis provides further support and suggests that many of the large-scale tourism activities and festivals planned by PAHSMA threatened (or indeed damaged) the physical integrity of the ruins.

While greater emphasis has been placed on conservation, some of the 'pay-to-enter' and 'gothic horror' tourism elements are still present today. For example, plays were offered in the Separate Prison in 2020 to provide tourists with an 'exclusive opportunity' (Parliament of Tasmania, 2020, 15). Linking this to Stone's (2006) classification of dark tourism sites, it is apparent that Port Arthur, under Jackman's description falls within the 'dark fun factory' category:

A Dark Fun Factory alludes to those visitor sites, attractions and tours which predominately have an entertainment focus and commercial ethic, and which present real or fictional death and macabre events. Indeed, these types of products possess a high degree of tourism infrastructure, are purposeful and are in essence "fun-centric". (Stone, 2006, 152)

One element that remains of the 'dark fun factory' at Port Arthur, is that, like Stone's description of the London Dungeon, the site 'offers a socially acceptable environment', that is 'highly visual' and 'family-friendly' that enables tourists 'to gaze upon simulated death and associated suffering' (Stone, 2006, 153). Yet, the site also incorporates features of a 'dark dungeon', namely, it presents 'bygone penal and justice codes to the present day consumer, and revolve around (former) prisons and courthouses' (Stone, 2006, 154). While entertainment remains a large part of these dark tourism destinations, there is also a focus on education, a 'relatively high degree commercialism and tourism infrastructure' (Stone, 2006, 154). Even the historical tours of Port Arthur, while lacking the high degree of tourism infrastructure focused on frivolity and entertainment (with a dash of education), place the site on the 'lightest' side of Stone's dark tourism spectrum.

It is also important to note here that the convict 'narratives' are heavily drawn from the records of the criminal justice system:

Any reconstruction of the life of a prisoner that draws on information in court, police, prison, and convict records will thus necessarily be shaped by the way in which those institutions wished to portray the perceived threat posed by that individual. They will not necessarily reflect the prisoner's life as they or their family may have known it. (Maxwell-Stewart and Nicholson, 2017, 724)

As such, a tourists' understanding of the convict experience is based largely, and in many cases, solely, upon the records kept by criminal justice administrators. In some cases, there are details of some of the convicts' lives pre- and post-transportation, however, this information again usually relies upon records kept within government archives. This inevitably means that the nuances between the individual experiences of convicts remain centred upon their interactions with the criminal justice system, rather than exploring more of their life courses or details about the differences between being sent to a rural or urban punishment location.

The 'dark fun factory' and 'dark dungeon' elements of Port Arthur tours have been criticised since the 1930s. According to the *Evening News* (1939, 4), 'the horrors of Port Arthur convict settlement have been grossly exaggerated by people profiting from the sale of souvenirs to tourists'. Johnnis Dederick Danker, Hobart's picturesque historian, claimed that many of the 'souvenirs' bought at the site, such as the 'cat-o'-nine-tails used at the settlement' or the 'carvings by notorious convicts ... were made long after the settlement had been abandoned' and that 'Leg-irons of ridiculous weight are in existence' (Evening News, 1939, 4). The guides were accused of creating 'lies that have painted Port Arthur with horror and morbidity' (Evening News, 1939, 4) and providing 'bloodthirsty and bloodcurdling' tales to tourists (Mercury, 1939, 2). More recently, Goc (2002, 26) wrote that the Port Arthur management had 'commodified the sacred ruins of our ancestors into a tourist asset ... [and] its transformation [had] effectively disguised the pain, the suffering, the dark brutal years and [had] compounded a denial that still permeates Tasmania today'.

## Local economy

While it is easy to sit back and criticise the nature and degree of tourism at Port Arthur, and the inclusion of 'gothic' or 'fun-centric' entertainment, the reality is that the site of Port Arthur is situated in a rural location and has numerous competing demands placed upon it – from the local community, the government and tourists themselves. Port Arthur has undergone consistent changes to reflect the changing government ideologies as well as to ensure the ongoing survival of a rural community. The benefits of tourism to the local economy of Carnarvon were quickly established in the nineteenth century, and while the desires of the tourists did (and continue) to shape the type of tourism activities on offer, the site continues to entertain, educate and connect tourists of all ages and nationalities to the cultural convict heritage of Australia.

Yet, while the benefits of tourism have long been established, in 1983, Kay Daniels recognised the ongoing dilemma within Tasmania between the desire to shy away from the shame associated with convict heritage compared to the seemingly obvious economic potential to profit from convict heritage:

> Tasmania is a society which is still uncomfortable with its convict past, which sees its history as in some ways marked by a shameful inheritance. But Tasmania is also part of an economy which is making the past into a consumer item: through the media, the cinema, tourism. The past which is too awful to contemplate must be made to earn, to work, to create income and profits, to create employment. The great liability must be turned into an asset. What has been most hidden and most despised must be turned into a spectacle. Tasmania must make a cult of its past. (Daniels, 1983, 3)

While Daniels believed that tourism at Port Arthur could not rescue the state's economy, the clear economic rationale for developing and supporting tourism in the area has been clear from the start.

Turning again to the theme of historic tourism at Port Arthur, Jones (2016) highlighted different patterns in historic alcohol consumption at Port Arthur and attributes this to the rural location. For example, Jones (2016, 327) suggests that the artefacts found on the site suggest that, in the early to mid-1900s, while 'alcohol consumption was taking place [at the two places of accommodation] … other activities at the tourist site may have superseded alcohol consumption'. This is distinct from urban hotels where past research into nineteenth-century artefacts has found a high frequency of liquor bottles (Harris et al, 2004), as such, there is evidence to suggest that alcohol played a larger role in the tourism experience of those staying in the city throughout the nineteenth century (Jones, 2016). As there were (and remain) few sites of accommodation and places with a liquor licence at Port Arthur, this finding is not surprising – however, it does highlight some of the considerations that rural tourism destinations need to consider. For example, Corbin et al (2010) found that early park tourism to Yellowstone Park in the United States embodied a certain element of 'luxury', including the tourism companies supplying alcoholic beverages as part of a tourist's stay. This is reminiscent of the early trips to Port Arthur, where the feast and 'merriment' were an expected part of the long trip.

Today, while Port Arthur is more accessible, it remains a remote, rural area (PAHSMA, 2018) located approximately 100 kilometres from Hobart, and the boat trip there still takes between three and four hours (Mason et al, 2003). The Tasman Peninsula consists of 'remote, separate and sparsely populated' communities with a permanent population of approximately 2,000 people and the 'economy is characterised by a reliance on agriculture, fishing, forestry and tourism' (PAHSMA, 2009, 39). The remoteness of the site 'presents ongoing challenges in recruiting, retaining and accommodating employees' (PAHSMA, 2018, 19). Yet, it focuses on providing as much employment to the region as possible. For example, PAHSMA employs approximately 150 staff and is the largest employer on the Tasman Peninsula (PAHSMA, 2009) and 'actively sources both local produce and services ... resulting in a greater quality of food, improved local relationships and a reduction in food miles and travel time' (PAHSMA, 2018, 12). The site also acts as a local centre for the community, with the hosting of the annual Boxing Day Woodchop, memorial cricket match and Carols by Candlelight. In addition, local residents are offered free entry to the site and local artists are featured and sold at the site (both in the gift shop and restaurants).

Despite the high level of tourism activity at Port Arthur, the site does not 'generate sufficient income from its tourism operations to meet these substantial requirements. This creates an ongoing reliance on external funding' (PAHSMA, 2018, 19). As such, the PAHSMA is required to balance protecting the heritage values of the site (including conservation and interpretation) with ever-changing tourism markets and political ideologies to ensure ongoing tourism, as well as external funding support. In addition, due to its remoteness, 'Port Arthur must attract the visitor for a full day or two' (Staples, 1995, 38). This essentially means that tourism infrastructure and activities must cater to the desires of the tourist, again reinforcing the need for the site to sit on Stone's lighter side of the dark tourism supply spectrum and to incorporate elements of a 'dark fun factory' and a 'dark dungeon'. As such, local economies and tourism at Port Arthur have inevitably been shaped by its rural location.

## Rural idyll tainted with dystopia

One of the many attractions of Port Arthur is its natural beauty (see Figure 3.1). Before it became a prison, Van Diemen's Land was seen as a 'romantic vista, the magnificent panorama, the picturesque nature of an undulating landscape' (Daniels, 1983, 4). After it became a penal settlement, the natural beauty was maintained, and many, if not most tourists commented on the beauty of the natural landscape (the idyll), and the juxtaposition of the 'barbarous cruelty' (Leader, 1890, 5) inflicted on convicts at the penal settlement (dystopia):

Steaming slowly into the narrow inlet known as Port Arthur, it is difficult to believe that we are at the very threshold of a spot disfigured by one of the darkest blots in colonial history. All that is visible is a calm sunlight bay surrounded by luxuriantly wooded shores, sloping gently down to the water's edge. To associate the nameless horrors of conviction with such a scene seems a scarcely possible feat of mental gymnastics ... the forbidding outline of these prison buildings was not sufficiently recognisable to mar the beauty of the picture. Only afterwards did the majority of the tourists learn that the fairy-looking inlet which had aroused their admiration was Dead Island, literally packed with the bones of hundreds of wretched convicts, whose release had only come with death. (Argus, 1890, 6)

Yet despite Carnarvon's old world look and air of slumberous peace, there is a sense of something jarring and incongruous. ... There is nothing picturesque about these fallen barracks, nothing pleasant in their memory. They were hideous in their prime, and their ruins resemble the skeletons of some misshapen monster. (Ace, 1888, 4)

Even during the early stages of tourism, the convict establishment took on evil and dystopian qualities. One tourist claimed that they could only enjoy the beauty because they were 'free', insinuating that the convicts would not have found Port Arthur picturesque (Leader, 1890, 5); others noted the sounds of nature they could hear such as the waves breaking on the shore and how the 'sighing wind ceaselessly sobs a sorrowful dirge' (Grapho, 1925, 12).

Over time, natural disasters have affected the site, particularly during periods when there was little to no formal protection. These natural disasters have played a unique role in transforming the idyll into a dystopia. The bushfire season of 1897–8 caused tourism to briefly decline and, for a time, altered the natural beauty of the landscape. However, within days of the

**Figure 3.1:** The picturesque nature of the Port Arthur Historic Site

Note: The ruins of the penal settlement stand in dystopic contrast to the beatific landscaping.

Source: Peters, J. (nd), 'Historic Convict Structures in Port Arthur, Tasmania, Australia'

penitentiary being burnt, tourists again arrived to 'inspect the fire-guttered buildings, wander through the ivy-covered church ruin, in which daises were flourishing, and taking boat trips to the Isle of the Dead' (Weidenhofer, 1990, 128). The resulting burnt ruins became part of the focus of visitor accounts, commenting that it was still 'roofless and floorless, and begrimed with smoke. The coral plant creeps on the interior walls, and the fern flourishes where the floors have been' (Mercury, 1900, 3). One tourist depicted the burnt ruins as 'gaunt and skeleton-like' standing 'witness to the past' (World, 1919, 3), adding to the notion of rural dystopia and horror.

Over time, the dystopian element has waned (particularly to international tourists who may not have much knowledge of Australia's convict heritage) and been replaced with notions of romanticism. Like many other rural places, Tasmania has been romanticised as being characterised by 'rural innocence, of bush virtues, of the robust outdoors, the contemplative cults of man in the wild of which the cult of wilderness is the apotheosis' (Daniels, 1983, 4). This romanticism has surrounded Port Arthur, and as Daniels (1983, 4) argues, Port Arthur's convict 'stain' has been 'assimilated into the imagination' and romanticised 'by way of the cult of the romantic ruin, the cult of man as a figure in a romantic landscape'. This romanticism overlay has certainly become more apparent as time has gone by. As White has commented, for many, the initial impression of Port Arthur after its closure was one of horror and distaste. However, as tastes changed, the site became 'romantically wild, awe-inspiring and picturesque' and seen as a 'romantically picturesque ruin' (White, 2016, paras. 14, 16). PAHSMA (2019, para. 4) recognises that Port Arthur's 'landscape, ruins and formal layout symbolise a transformation in Australian attitudes from revulsion at the hated stain to honouring of and interest in the convict past'.

Port Arthur, being located in a rural setting, provides tourists with this unique juxtaposition that is often missing in urban areas. Aside from the early attempt to destroy Port Arthur and all memories of the site, there has been no pressure to rebuild or reinhabit the area, unlike many sites in urban settings. The remote setting enables tourists to feel the original isolation of the penal settlement, with few 'modern' sounds being heard at the site (traffic noise etc), thus further fuelling the romantic notion of the site. As such, Port Arthur remains a site of picturesque ruined buildings that 'project an idealised notion of rustic serenity contrasting dramatically with Port Arthur's penal history' (PAHSMA, 2009, 49).

While tourists have found this juxtaposition jarring and may have assumed that the 'beautification' of the site was for the benefit of distancing the site from its past horrors (or for making the experience more pleasant for tourists), the reality is that the parkland landscaping:

is the product of a much misunderstood aspect of the system of authority exerted over both convicted and free persons. The original gardenesque landscape was intended to symbolise for all inhabitants the desired qualities of a thriving society – order, discipline, beauty and obedience. The present landscape contains elements of the original penal landscape design, but over time has been modified to reflect both natural change and to facilitate landscape management. (PAHSMA, 2009, 49)

Owing to the extensive conservation of not only the convict heritage but also the building of Carnarvon and tourism activities, the PAHS is able to vividly convey to tourists the importance of the landscape and location of the site. Indeed, the landscape itself is one of the major 'displays' of the site and it:

reveals the creation and arrangement of functions within the space, and tells of the complex interplay between the natural environment and human activities and cultural perceptions. … Some aspects of the presentation of the landscape have been consciously created to enhance the appeal of Port Arthur to visitors, and also to contain and sanitise its powerful and confronting meanings. Because this aesthetic appeal is derived from an interplay between significant designed elements, natural qualities and park management practices, there are complex tensions arising between these aspects and other heritage values, such as the meanings associated with Port Arthur's incarceration and industrial functions during the convict period, and the rural subdivision pattern and character of the township period. (Context Pty Ltd et al, 2002, 58)

As such, the presentation of Port Arthur purposely maintains the juxtaposing images of idyll and dystopia because it ultimately represents both these conceptions. Further, the ruins are still considered gothic and invoke 'dark' themes, and the 1996 massacre added a further, and much more recent dystopian element to the site.

## Conclusion

The convict is an iconic symbol within Australia's cultural heritage. When one thinks of convicts, images of rural landscapes, rural pioneers and lawbreakers abound. While there are several major convict heritage sites located in urban areas (for example, Hyde Park Barracks and Cockatoo Island in Sydney, Fremantle Prison in Fremantle, and the Female Cascades Factory in Hobart), almost half of the convict sites listed on the World Heritage list are located in rural and regional parts of Australia and form a substantial part of Australia's domestic, and international, tourism market.

PAHS is one of Australia's most prominent convict tourism sites even though only 8.7 per cent of the 1,123 men that landed in Van Diemen's

Land between 1840 and 1853 spent time at Port Arthur (Maxwell-Stewart and Nicholson, 2017). A possible explanation for the popularity of the site, despite its low levels of convict incarceration, is that the PAHS offers tourists an informative, fun and memorable cultural institution that celebrates (and, to an extent, challenges) the dark side of Australia's convict heritage. While historically there were perhaps more 'dark' exhibits (many of which were pilfered by tourists) and encounters (for example, with ex-inmates), the site continues to engage visitors with uncomfortable and 'dark' exhibits and themes. Despite the dark themes, the site remains firmly on Stone's lightest side of the dark tourism spectrum, and the site is best described as a combination of a 'dark fun factory' and a 'dark dungeon'. The location, through its geographical isolation, demands that the management of the site cater to a range of tourists (in terms of age, mobility and language accessibility) to remain viable and attract people to its remote location.

Yet, while Port Arthur has always economically benefited the local community, it is important to recognise that tourism has not always been welcomed and that the community has suffered because of the convict 'stain'. The attempted distancing from the penal settlement through the renaming of the town to Carnarvon stemmed from the desire of a small community to forget the 'dark', painful and shameful past. Yet, tourism interest trumped those desires and indeed superimposed its own agenda onto the community. The community was forced to readapt and embrace the perceived benefits of tourism to their area. As such, there has been an ongoing negotiation, and renegotiation, of how the local community can balance the desire for tourism against their economic, cultural and social life.

For the Port Arthur community, in some ways, the eventual acceptance, and later embracement, of convict tourism has certainly paid off. Port Arthur provides the community with the largest income provider in the district – the site employs local people, as well as sells local products in the gift shop and sources local ingredients for the restaurant. However, in other ways, these same benefits can hurt the community. For example, during the COVID-19 pandemic when the site was shut, the loss of local revenue impacted the whole community. The area is also intrinsically linked with notions of 'convictism', making it difficult to create new narratives that emphasise other aspects of the region's history, as the residents of Carnarvon discovered. As such, future generations have been enlisted in the ongoing convict narrative that shapes their cultural life. Further, community members continue to suffer from the 'shame' of the 'convict stain', impacting their social life and how Tasmanian-born residents are treated more broadly across the Australian community. As Tumarkin (2005, 6–8 'It Goes On') has sensationally written, in the not-too-distant past, many Tasmanian people have been stigmatised and the state has been portrayed as a place populated by 'stupid, rednecked, inbred, racist, homophobic, two-headed convict bastards'. These same stereotypes have certainly never been applied to Sydney, despite Sydney being the first penal colony of Australia.

Such portrayals fit in with the narrative of Port Arthur as a rural dystopia: a place tainted by the barbarous cruelty and nameless horrors inflicted on the original inhabitants, and lasting ruins that resemble skeletons of misshapen monsters. At the same time, the location provides a romanticised rural setting, where the dystopian element can be moderated into a tolerable and palatable tourist encounter.

# 4

# Bushrangers

Nineteenth-century outlaws (those who live 'outside of the law'), highwaymen or bandits were known as 'bushrangers' in Australia. The *Sydney Gazette* described bushrangers in 1805 as 'a group of suspected highway robbers, possibly escaped convicts, who often waylaid travellers in the bush' (cited in Tranter and Donoghue, 2008, 374). The myth of 'bushrangers' holds a strong place in Australia's national identity and is often featured within popular culture. Many Australian rural/regional towns have a close affiliation with their bushranger past and have built tourist attractions showcasing the lives and exploits of these personalities. According to Australian historian Russell Ward, the 'convict system manufactured bushrangers' (2003, 147), and nearly all of the early bushrangers were prior convicts. With the cessation of transportation to NSW in 1840 (and transportation ceased everywhere to Australia in 1868), bushrangers began to replace convicts as Australia's romanticised lawbreakers, continuing the focus on rural Australia as housing violent and dangerous criminals.

Bushranger tourism sites often fall on the lighter side of Stone's (2006) spectrum, with a focus on providing family-friendly commercialised experiences. Some sites could be likened to Stone's (2006) categorisation as 'dark fun factories', which provide 'fun-centric' experiences with high levels of tourism infrastructure. Sites focusing on bushrangers can often be perceived as less authentic because the violence and crimes committed by bushrangers are presented in a highly sanitised manner. Sites with less tourism infrastructure may constitute 'dark conflict sites' where, although the focus is not on 'warfare', the site does represent a violent conflict or a location associated with a violent individual, such as a bushranger. By featuring bushrangers that are often romanticised, the 'narrative' of sites diminishes both the 'authenticity' and 'darkness' of the site.

Dark conflict sites can also play an important role in creating storyscapes of bushranger activity across a broad geographical location, encompassing several towns and thus providing tourism opportunities for a region, rather than one specific town. In essence, this configuration of attractions and towns are 'bound together by sharing a theme which tells a story', becoming what Rodaway calls a 'themescape' (Fagence, 2017, 452). Fagence (2017, 455–456) argues that 'themescapes' promote the 'theme' over the 'route', and that perhaps a better term could be a 'trailscape', which 'may be conceptualized as a specially contrived mode of story-telling' that relies on specific geographical locations across a range of sites 'facilitated by a convenient trail' to create a broader narrative.

Whether it is themescape or trailscape tourism, bushranger tourism also allows some sites to have more tourism infrastructure while allowing other locations in-between these larger tourism destinations to remain as dark conflict sites without the fun-centric, commercialisation focus. The varying approaches to such sites add more 'authenticity' to the overall tourism experience of encountering representations of bushrangers in rural Australia, and allow tourists the opportunity to experience the 'dark' and 'light', or just one or the other depending on motivations and expectations.

Two well-known bushrangers, Ned Kelly and Captain Thunderbolt (Frederick Ward) are explored throughout this chapter to illustrate long-established dark tourism sites available in rural and regional Australia. According to a survey conducted in 2005, 80.4 per cent of respondents knew about Ned Kelly, and the third most well-known bushranger was Captain Thunderbolt at 12.4 per cent (Tranter and Donoghue, 2008). Within Australia, bushrangers tend to be celebrated in similar ways to international 'outlaws' such as Robin Hood (United Kingdom) and Jesse James (United States).

## Ned Kelly

Ned Kelly has been described as a 'notorious bushranger, popular icon and national identity. ... Today, Ned Kelly is indelibly stamped on the nation's psyche – part villain, part folk hero, but also a man whose courage and defiance is uniquely Australian' (Visit Melbourne, 2022, paras. 1, 7). Edward (Ned) Kelly was born in Beveridge, rural Victoria, in 1854 and experienced a tumultuous early life, including the death of his ex-convict father when he was 12 years old. From the age of 14, Ned developed a colourful criminal career, including assault, participating in highway robbery while being an apprentice to the Victorian bushranger Harry Power, feloniously receiving a horse, resisting arrest, and humiliating and assaulting a police officer. In 1878, Kelly formed his own 'gang', which consisted of (mainly) Ned, his brother Dan and friends Joe Byrne and Steve Hart, although they received help from other family members and friends. Throughout the gang's endeavours, they attracted some local appreciation, and for over 130 years, Ned Kelly has been referred to as Australia's Robin Hood – a 'larrikin' who was 'irreverent of authority and disdainful of conformity' and 'a champion' for small land holders 'who felt disenfranchised and victimized by bureaucracy and community policing' (Scott and MacFarlane, 2014, 717).

Owing to the rough and isolated rural terrain along the border of northeast Victoria and southern NSW, there was limited policing of the area and the Kelly Gang was able to steal and redistribute livestock with impunity. However, the Kelly Gang became targeted by police efforts after they shot a police officer in the wrist and relocated to bushland to hide. Police were dispatched to Stringybark Creek to look for them, and in a subsequent encounter on 25 October 1878, the Kelly Gang killed three police officers: Sergeant Michael Kennedy and Constables Thomas Lonigan and Michael Scanlan. A fourth police officer, Constable Thomas McIntyre, escaped. Kelly's version of events was that the gang was acting in self-defence and that he let Constable McIntyre escape. However, the forensic evidence, and eyewitness accounts from McIntyre, indicate that the gang shot the mounted officers before they could surrender or draw their weapons (Scott and MacFarlane, 2014).

Two days before the most famous of the Kelly Gang's sieges at Glenrowan, Joe Byrne shot and killed Aaron Sherritt (Scott and MacFarlane, 2014). According to McDonald and Davies (2015), the murder was an act of retribution and was also timed to attract police attention – the outlaws prepared to ambush the police response by cutting the telegraph wires in Glenrowan and forcing railway workers to dismantle part of the railway track that would inevitably carry police reinforcements to the town. The intention was to kill all police officers while they struggled to disembark from the derailed train. After damaging the tracks, the gang took 60 hostages into the nearby Glenrowan Hotel (also known as Ann Jones's hotel) to await the police train. Kelly allowed a hostage to leave, Thomas Curnow, who was able to alert the police and prevent the derailment of the train.

At 3 am on 28 June 1880, the police surrounded the Glenrowan Hotel while the four gang members, protected by 'crudely fashioned quarter-inch thick metal armour ... fired upon the advancing police from the hotel' (Scott and MacFarlane, 2014, 727). The armour, which was to become a famous symbol of Ned Kelly, weighed 90 pounds (Visit Melbourne, 2022). In total, more than 50 police officers were involved in the siege. The police shot Kelly in his legs (the one part of his body unprotected from the armour) after he left the hotel, and, incapacitated, Kelly was subsequently arrested. Joe Byrne, Dan Kelly and Steve Hart, along with three civilians, died throughout the siege. The siege itself also attracted a large crowd of onlookers, which provides us with a rare example of what we might call 'live dark tourism' (that is, eager spectators to the events as they unfolded).

A preliminary court hearing for Kelly was held at Beechworth Courthouse in August 1880, with the trial occurring in Melbourne in October 1880, where he was found guilty and sentenced to death. On 11 November 1880, Ned Kelly was hung at the Old Melbourne Gaol at the age of 25. Shortly after his death, several plaster death masks were made; one went on public display the next day in Bourke Street in Melbourne (Scott and MacFarlane, 2014). His body was buried in the gaol yard, but in 1929, the remains were transferred to a mass grave at Pentridge Prison in Melbourne. During the exhumation of Kelly's grave (and others), souvenir hunters or 'crowds of people swarmed from all directions, grabbing whatever they could get a hold of', which included the bones of Kelly and other deceased inmates (Priestland, 2018, para. 11). The contractor retrieved the skull that was believed to have been Kelly's before others could take it. However, forensic analysis in 2009 revealed that the skull was not that of Ned Kelly. In 2009, the skeletons buried at Pentridge Prison were exhumed, and DNA testing proved in 2011 that one of these skeletons belonged to Kelly. After testing, his remains were later buried in an unmarked grave (to deter vandals) at a small graveyard in northeast Victoria. In September 2023, reporter James Phelps announced that new forensic testing confirmed that Ned Kelly's skull had been buried (in multiple parts) in his burial box at Pentridge Prison with the rest of his remains.

From his death to the current day, Ned Kelly has been used to promote tourism, products and events across Australia. For example, in the opening ceremony of the Sydney 2000 Olympic Games, 'a group of armoured Kelly figures paraded around waving mock firearms spouting streams of sparks' (Tranter and Donoghue, 2010, 190). Merchandise utilising the Ned Kelly image is available across Australia: including, soap, towels, coffee mugs, replicas of his armour and even university criminology textbooks. The iron helmet Ned wore in the last siege was presented around Australia in 2006 in an exhibition of National Treasures (Tranter and Donoghue, 2008). In 2007, the Australian Survey of Social Attitudes (AuSSA) found that the majority of Australians (57 per cent) believed Ned Kelly is important as a symbol of Australian identity.

In more recent years, scholars and the relatives of the police victims have been calling for a reimagining of the story of Ned Kelly and his gang at tourism destinations and across the wider popular culture narrative. As Scott and MacFarlane (2014, 716) adequately summarise, 'a close examination of his developmental history and subsequent criminal behaviour reveals that Kelly was a violent and vindictive man who demonstrated prominent psychopathic features including pathological lying, callous lack of empathy for others and a parasitic lifestyle'. As such, Ned Kelly is one 'of Australia's most divisive bushranger stories' (Somerville, 2017, para. 15). Despite this, there is clear tourist and souvenir looting/hunting interest in Ned Kelly and his gang. As already noted, the site of his grave was looted, and his .31 Colt Pocket revolver (thought to have been used to shoot the officers at

Stringybark Creek and Glenrowan) was stolen in 1976 from an Australian historical exhibition in Chicago, United States (Canberra Times, 1976). With the ongoing fascination and reverence of Kelly, it is unsurprising that Ned Kelly has become a key tourist icon in rural Victoria and NSW.

## Ned Kelly touring route

As the story of Ned Kelly can be told in multiple locations across two different states, travelling 311 kilometres from Jerilderie (NSW) to Melbourne (Victoria), tourists can immerse themselves in the story for several days and provide economic support for multiple locations in rural and regional towns of NSW and Victoria. Taking advantage of this, a 'touring route' has been established, with its own website to guide tourists along 14 significant locations associated with the bushranger's infamous story. The collective approach to guiding tourists along Ned Kelly's story ensures that each town has its own 'story' while supporting and encouraging tourism to other similar locations. The 'selling' of Ned Kelly and his gang extends to a range of merchandise available at multiple locations along the touring route, with the range of merchandise attracting derision from some journalists: 'You can buy Ned Kelly T-shirts, stubby [small beer bottle] holders, playing cards, moneyboxes (seriously? He was a robber), pies, vanilla slices and baseball caps (he didn't play baseball)' (Silvester, 2018, para. 9).

Towns along the route differ in size, from Eldorado and the Woolshed Valley (Victoria, population 385) to Melbourne (Victoria, population 5.078 million). While not all sites along the touring route are associated with the criminal activities of Ned Kelly and his gang, most are. Some towns specialise in Ned's early life, while others focus on Ned's apprenticeship with bushranger Harry Power. While not focusing on specific stories of violence and death, such sites provide further context to the Ned Kelly narrative and provide tourists with multiple stopping points to support local economies. While there are three main 'crime-related' sites along the touring route (to be explored in detail in what follows), numerous other towns provide insight into the 'Kelly Gang manhunt'. For example, the Victorian town of Benalla provides tourists with the opportunity to explore some of the police and court cells where Kelly was held. Further, the Benalla Costume and Kelly Museum offers tourists a view into the violence and pain of the gang's last siege by displaying Ned Kelly's bloodstained sash that he wore under his armour during the final siege. Eldorado and the Woolshed Valley offer tourists the opportunity to visit a 'murder scene' (where Joe Byrne, lieutenant of the Kelly Gang shot his friend Aaron Sherrett), the Kelly Cave (where the gang briefly lived after the police killings) and the 'police caves' where police camped to watch for movements of the Kelly Gang (Ned Kelly Touring Route, 2022).

One of the few sites along the route to focus on the victims of the Kelly Gang is Mansfield (Victoria, population 4,787). Here, tourists can visit the cemetery to find the graves of the three police members murdered by the Kelly Gang at Stringybark Creek, the operational Mansfield Courthouse where Ned was proclaimed an outlaw and a police memorial in the form of a marble Troopers' monument (funded by public donation), which is featured in a roundabout in the centre of the township, honouring the three murdered Victoria Police members.

### Stringybark Creek

Tourism to Stringybark Creek began shortly after the execution of Ned Kelly in 1880 and continues to be a popular stop along the touring route, attracting 20,000 visitors annually who are motivated to witness 'a dark chapter of Australian history' (Somerville, 2017, para. 3). Some tourists are 'Kelly fans', others are related to the police officers or the Kelly Gang members, while others seek to understand and/or commemorate the lives of those involved. The site is isolated, difficult to find and is essentially bushland:

> Stringybark Creek Historical Reserve, 270 kilometres from Melbourne, is a pretty piece of bush that would attract the occasional fisherman, yabbyist, nature lover, trekker, or hitman looking for the ideal spot to dig a shallow grave, if it was not for the fact that it contains a location that has become part of the very fabric of Australia's heritage. (Silvester, 2018, para. 4)

A reoccurring theme throughout this book is the juxtaposition of the rural beauty with the monstrous acts that occur in these places. As one traveller has written of the Stringybark Creek Historical Reserve, 'It's hard to believe this very pleasant spot was the location of one of the most notorious killings in Victorian history' (Coia, 2012, para. 20).

Stringybark Creek has had a contentious history of providing tourists with authentic narratives. The first tourist marker of the site was the 'Police Tree', later known as the 'Kelly Tree', which originally started as an impromptu memorial to the police officers but later became overshadowed by the 'hero' legend of Ned Kelly and tourism opportunities. The tree is one of the main 'attractions' for tourists with rumours that it contains a bullet from the Kelly Gang's shoot-out with the police (Whitley, 2020). However, this 'tree' has been replaced at least twice, once in the 1920s and again in the 1930s. As such, the 'tree' is an unauthentic replica. This last tree became 'a living memorial to the three police troopers who lost their lives there, with their names carved into the trunk' (signage at the Kelly Tree cited in Coia, 2012).

In 1985, a metal plaque shaped as Ned Kelly's helmet 'was affixed over the top of the remaining visible name – that of Constable Lonigan' ('mementos, monuments and marked trees' sign 2018, located at Stringybark Creek Historical Reserve) causing indignation from the descendants of the victims. There has since been speculation that this tree (and perhaps even the original tree) was not located near the place of the shootings, which has led one commentator to state 'should you come to Stringybark Creek, you'll be served up a whole heap of tosh' (Whitley, 2020, para. 11).

On 26 October 2001, an official police memorial was added to the area with an additional plaque honouring the three police victims placed on a prominent rock at the picnic site to commemorate the anniversary of the murders. In 2007, a story appeared in the *Border Mail* detailing past vandalism and then the theft of the plaque. The local police reported that the theft was 'a slap in the face' and blamed souvenir hunters for its theft (Bunn, 2007, para. 6). In 2009, further work was added to the area to make it more accessible for tourists. Walking tracks, picnic tables, fireplaces and camping grounds were either built or upgraded, and new signs were erected to tell the story of the shootings (Somerville, 2017).

As a result of these signs, the site came under criticism for glorifying Ned Kelly, overlooking the victims and, in essence, being used to promote Ned Kelly tourism industries, including the Ned Kelly Touring Route whose text appeared on the signs at the Reserve. As journalist John Silvester (2018) points out, the touring route only provided the last names of the police officers in their recount of the murders, while affording the murderer his first name. In addition, the sign downplayed the brutality of the murders (including evidence to suggest that one of the policemen was shot while surrendering), as well as the clear intent to commit murder. In essence, the 'memorials' to the police officers were overshadowed by the desire to 'sell' Ned Kelly as a hero and legend. For example, the path to the area 'was lined with little Ned Kelly helmets as guide markers' (Silvester, 2018, para. 28).

As a consequence of these criticisms, a new police memorial was designed and rededicated on 8 December 2018. As part of the new memorial, the old 2001 memorial plaque on the rock was removed, the picnic area was relocated and new memorial boards advising visitors about the lives of the police who died were erected to ensure that they were 'no longer … bit players in the fake Kelly legacy' (Silvester, 2018, para. 36). An updated 1.4-kilometre walking track takes tourists from the memorial to the Kelly Tree and back, following the path of Constable McIntyre's escape. One of the 'features' of this path is that it also follows the trail of gunfire that targeted and, ultimately killed, Sergeant Kennedy. The walk takes approximately 30 minutes and undoubtedly follows a 'dark path' that was once taken by two men fleeing for their lives. At various points along the track, visitors can read memorial plinths for each of the fallen officers that detail the event as well as their personal and police service history. In addition, 'Interpretive signs around the reserve contain pictures and newspaper clippings from 1878, telling the story of what happened at the infamous shootout 140 years ago' (Strahan, 2018, para. 28), and tourists are told that the exact sites of where the police camped or were killed are unknown, providing more authenticity to the site.

Tourists also have the opportunity to visit Stringybark Creek as part of a helicopter 'Bushranger Tour', operated through the Alpine Helicopter Charter company. For $380 per person, twin share, tourists can be flown to 'Stringybark Creek where the siege took place. On your walk through this native bushland it's easy to imagine the sounds of gunshot and the echoes through the ranges all those years ago' (Alpine Helicopter Charter, 2021, para. 2). Like other 'dark' tours, this tour emphasises the need to experience the location first-hand so that tourists can imagine gunshot echoes. However, the option of taking a helicopter means that tourists can dally in the rural environment for a short period, remaining based in the comfortable urban environment.

## Glenrowan

Glenrowan is a small town approximately 240 kilometres northeast of Melbourne, Victoria, and has a population of 1,049 people. The Glenrowan township 'is the gatekeeper of the "Ned Kelly – the bushranger" history' (Rural City of Wangaratta, 2022). Since the siege at Glenrowan, the town has become synonymous with the famous bushranger, and the town's tourism strategy relies on Ned Kelly's final police siege to attract tourists to the town. While there are 49 accommodation establishments in the wider area (Wangaratta, Benalla, Milawa etc), only one is in Glenrowan itself and is named with reference to the bushranger legend – the Kelly Country Motel, providing tourists a complete immersion experience in the Kelly trailscape. The Glenrowan Heritage Precinct was recognised as having national significance when it was added to the National Heritage list in 2005.

The town has structured itself to preserve and commercialise the final siege, and the heritage precinct includes an eight-hectare site incorporating the original railway platform, the siege site and the location of the inn. One of the main tourist attractions is a 6-metre-high 'big' Ned Kelly statue in the main street to attract tourists (see Figure 4.1), which is very much in the Australian tradition of 'big things'. As Nichols and Freeman (2021) have argued, Australian local businesses and governments have created 'big' statues of local industries or iconic people/animals/inanimate icons in order to attract tourists. In other words, the big Ned Kelly is the result of the local area branding or commercialising the outlaw in a kitsch manner, or what Nichols and Freeman (2021, para. 12) would term a 'brandscape'.

**Figure 4.1:** The six-metre-high 'big' Ned Kelly statue, Glenrowan, Victoria, Australia

Source: Fletcher, R. (nd), 'Giant effigy of Ned Kelly outside Kate Kelly's Tea House Glenrowan Victoria Australia'

There are numerous Ned Kelly replicas around the town, with several standing 'guard' over a local café and souvenir shop (the Glenrowan Vintage Hall Café – all four of the Kelly Gang stand in armour on the balcony watching over the street, providing a fantastic photo opportunity for tourists to stand with the 'gang'), and one sitting and resting against a log where Ned Kelly was captured by police. There are various markers and signposts throughout Glenrowan informing tourists of the importance of each of the sites and creating a flowing narrative throughout the town (and wider trailing route).

Within the Glenrowan tourist centre, tourists are offered a 40-minute animatronic theatrical retelling of the siege and last stand utilising 'the brilliance of live special effects and animatronic figures' to 'transport' visitors back to 28 June 1880 (Glenrowan Tourist Centre, 2022, para. 1). According to the *Sydney Morning Herald*:

The Ned Experience was lampooned by comedians Mick Molloy and Tony Martin in the 1990s on ABC *TV's The Late Show* and more recently has polarised social media travel websites. It's still here, and when we went we too wondered what the history of the Kelly Gang has to do with a Vincent Price-style introductory monologue presented by a talking face projected onto a robot, pumpkins, urinating robots and a giant Ned Kelly. It's an immersive experience that blends Hammer Horror and Disneyland with a brand of kitsch all of its own. For that it is worth a visit. (Cornish, 2016, para. 5)

Tickets to the show start from $15 for primary school children to $32 for adults. The tourist centre also houses its own gallery and museum and sells Kelly souvenirs.

Kate's Cottage, Gifts and Souvenirs shop, named after Ned's sister, is located opposite the 'Big Ned Kelly', and provides access to a replica of the former Kelly homestead, where tourists can experience the 'stark reality of the hardships that our forefathers had to endure as this colony was being settled' (Kate's Cottage, 2022a, para. 2). The museum displays numerous replica artefacts including the legendary armour, one of the death masks (and related phrenology information) as well as numerous photographic representations of the era. To access the homestead and museum, tourists must first enter the gift and souvenir shop, which according to the website, 'proudly stocks the LARGEST range of NED KELLY SOUVENIRS' including clothing, armour (as well as children's helmets), and charms' (Kate's Cottage, 2022b, para. 3). Reflecting the lightness and family-friendliness of the interpretation of Ned Kelly as a tourism gimmick, the merchandise associated with Ned Kelly is extremely light-hearted: 'in an attitude reflecting a post-modern approach to tourism, the Glenrowan community advertises and promotes its souvenir as gimmicks, a mockery of the souvenir selling process' (Pearce et al, 2003, 79).

It is clear that Glenrowan has adopted Ned Kelly as the 'hero for their marketing, theming, or branding', emphasising the bushranger legend in food, souvenirs and accommodation establishments (Pearce et al, 2003, 74). As Staples (1995, 38) commented, the focus on Ned Kelly has served as 'a device to differentiate itself and attract some traffic from the Hume [Highway] to buy food and petrol'. As such, Glenrowan has successfully captured a day tourism market (and possibly overnight with the Kelly accommodation), which connects with a wider 'Ned Kelly' narrative that spans the 311 kilometres through NSW and Victoria rural and regional towns through to Melbourne.

## Beechworth

Beechworth is a popular tourist destination because of the beautiful historic nature of the town and, importantly, for its strong 'connection' to the Ned Kelly story. Beechworth previously held an annual event where tourists (and locals) could re-enact the trial of Ned Kelly in the Historic Beechworth Courthouse. This later turned into a three-day event, and one year, an inn was burnt down as part of the historic re-enactment. However, after more than 15 years it was replaced by 'A Gold themed' weekend in 2017, focusing on the history of the Gold Fields.

The Beechworth Courthouse has been preserved, offering tourists an 'authentic' opportunity to explore the original furniture and fittings that Ned would have experienced while standing trial. To further immerse visitors, the Courthouse offers 'an atmospheric soundscape' featuring expert historians bringing 'to life some of the dramatic events which took place inside these walls, including the committal trial which sealed Ned Kelly's fate' (Indigo Shire Council, 2017a, para. 1). The guides to the site, from the Historic and Cultural Precinct team are on hand to provide stories and details of the Kelly family, and school groups can exclusively book the Courthouse to re-enact famous trials. The Courthouse received a million-dollar upgrade from 2022 to 2023 to focus more attention on the Kelly narrative, including 'traditional museum displays, original artefacts, museum signage and modern projection technologies to allow visitors to step back in time and experience these significant events' (Indigo Shire Council, 2022a, para. 3).

In addition to the crime and justice type experiences, Beechworth houses the Robert O'Hara Burke Museum, erected in 1857, which displays gold rush relics and artefacts (authentic and replicas) of the Ned Kelly Gang. The Explore Beechworth website states that Ned Kelly's death mask is 'undeniably the most fascinating piece' in the museum (Indigo Shire Council, 2017b). The museum also offers tourists a mask that they can wear for a photo opportunity (Lonely Planet, 2022).

Beechworth's Historic and Cultural Precinct offers a daily 45-minute guided Ned Kelly walking tour. The tour leaves from the Visitor Information Centre and is relatively inexpensive at $5 for general admission. Tourists are taken around the town to see the buildings 'associated with the Kelly legend and hear stories behind the events which eventually brought the Kelly era to an end' (Indigo Shire Council, 2017c, para. 5). It is estimated that the entire Beechworth Historic Precinct attracts 25,000 visitors a year (Indigo Shire Council, 2022b), indicating that tourism provides substantial revenue for the town. While not all this tourism is based on Ned Kelly, it is a large focus, and being only three hours from Melbourne enables day tourists who want to experience more of the Kelly story in a regional setting.

## Melbourne

Despite Ned Kelly living and travelling in rural Victoria, the 'largest and most significant collections of original artefacts' are housed in Melbourne (State Government of Victoria, 2021), reinforcing the urban bias of housing and promoting authentic artefacts. Melbourne's most well-known connection to the Kelly story is based around his execution at the Old Melbourne Gaol on 11 November 1880, and his story features heavily at this site. Indeed, Witcomb (2013, 153) states that the Old Melbourne Gaol is 'famous as the site of Ned Kelly's hanging', and Welch (2013, 483) argues that the site is excessively flattering of Kelly, making his story 'rather hagiographic' by having its commercial logo featuring a drawing of Kelly's iconic helmet worn during the Glenrowan Siege (displayed on brochures, merchandise and, of course, the signs for the building). Ticket prices for adults are $33, and the site offers a range of organised night tours ('Ghosts? What Ghosts!' and 'Hangman's Night Tour') where specific 'dark and grisly stories' are told (Old Melbourne Gaol, 2022, para. 1).

As part of a wider dark tourism site, Ned Kelly's cell, a plaster cast of his 'death mask' and a replica suit of armour worn at Glenrowan during the last siege, are all on display at the Old Melbourne Gaol. The original scaffold, which was relocated to Pentridge in 1932 (and then transferred back to the Gaol in 1975), was reinstalled in the Gaol in 2000 – allowing tourists the opportunity to see the 'real' scaffold used to execute Ned Kelly. In the daytime, children are encouraged to 'dress up in Ned's armour!' for candid photographs; yet during the night 'hangman tour', visitors learn that the armour stands over the drainage site where the hanged people's last excreted bodily fluids were washed away, making this 'fun-centric' activity unknowingly dark.

The Victorian Police Museum (either free entry or a gold coin donation) displays Ned Kelly's armour, revolver and (bloodstained) cartridge bag. The narratives presented focus very much on the 'criminal' Ned Kelly and invoke empathy for the victims of the gang's criminal behaviour, particularly the police search party at Stringybark Creek. A quote from the surviving policeman from Stringybark Creek is prominently displayed on the wall near the armour of Dan Kelly and Steve Hart:

> Lonigan's body was visible from where I stood and I tried to keep myself from looking at it, lest it should unnerve me. ... The pallor of death had spread over his countenance, and the setting sun ... had cast the long shadows of the forest trees over his body. ('Thomas McIntyre' exhibit, Victorian Police Museum)

The quote focuses on the experiences of McIntyre being confronted with the death of his colleague in rural Victoria. The focus on the police officers and condemnation of Ned Kelly as a criminal is unsurprising for a police museum and offers an important narrative to the Kelly trailscape.

Across the sites in Melbourne, there is very little recognition or engagement with the rurality of the sites that the Kelly Gang encountered. In the recounts of the Kelly Gang exploits at the Old Melbourne Gaol, including the Glenrowan Siege, there is no information about the rurality of the townships, with just a brief mention that Kelly was born in 'country Victoria' (Early Years exhibit Old Melbourne Gaol). As such, the rural landscapes remain unexplored within these urban environments. Similarly, at the Victorian Police Museum, there is little reference to the rural nature of the events, with only occasional references such as Sergeant 'Kennedy was pursued for nearly a kilometre in a desperate attempt to stay alive. His body was discovered in rugged scrub five days after the attack' (Sergeant Kennedy's Murder exhibit Victorian Police Museum).

## Captain Thunderbolt

While some have questioned Ned Kelly's status as a 'bushranger' (because he did not steal from mail coaches or commit highway robbery), Frederick Wordsworth Ward, also known as Captain Thunderbolt, conducted numerous 'bushranging' activities across large rural and regional areas of NSW. Ward's first conviction for criminal activity occurred in 1856 when he was sentenced to Sydney's Cockatoo Island for horse stealing. Folklore narrates that Ward became known as Thunderbolt in December 1863 after he hammered on the door of a toll keeper he was robbing 'causing the bushranger's victim to proclaim: "By God, I thought it must have been a thunderbolt". It is claimed that Ward, his gun drawn, replied: "I am thunder and this is my bolt"' (Australian Government Sydney Harbour Federation Trust, 2021, para. 12).

His 'fame' is celebrated in the New England area of NSW, particularly around Uralla; however, his criminal activity, including robbing mail coaches and holding up innocent travellers, targeted a much broader area. Like Ned Kelly, Captain Thunderbolt has been romanticised because of his 'agreeable appearance and conversation' and his 'gentlemanly behaviour and avoidance of violence' (Crittenden, 1976, para. 5). In May 1870, the police became aware that Thunderbolt was in the Uralla area and a pursuit through the bush concluded with Constable Walker confronting and shooting Thunderbolt in Kentucky Creek. Thunderbolt died on 25 May 1870 while being transported back to Uralla. Hundreds of tourists 'flocked to see the body of Captain Thunderbolt (1835–1870) after his death and for a shilling, you could buy a postcard of his bullet-ridden body' (State Library of New South Wales, 2022, para. 2).

*Museums, memorials and natural locations*

The first main tourist attraction to be 'constructed' was Thunderbolt's gravesite, located within the local Uralla Pioneer Cemetery. The site was first marked with a wooden cross, however, this was replaced by a headstone in 1914, funded by local New England residents. In 1919, the site appeared to be attracting substantial numbers of tourists:

> There is a biscuit tin near the grave, and many are the different ideas of the man of the road expressed on papers enclosed in it by visitors to the tomb. People from all over Australia and visitors from America and the old country are among those who have left memos in Thunderbolt's tin. ... Thunderbolt's tomb, though not an imposing affair, and shaded only by a stunted bush, seems to be the Mecca for a number of pilgrims, and it has been the subject of innumerable photographs. (Uralla Times and District Advocate, 1919, 2)

Twenty years later, the *Western Mail* (Perth) printed the reminiscences of a tourist who had found a tin box next to the grave with numerous cards and papers left by other visitors in 1929 (Spero, 1939).

In 1988, during Australia's Bicentenary, the township of Uralla (population of 2,743) unveiled a $70,000 bronze statue of Fred Ward (Captain Thunderbolt) on the main street (which acts as the main inland highway between Sydney and Brisbane) (see Figure 4.2). According to the press at the time, the community was split in opinion over the erection of the statue, with neighbours not speaking to one another and businesspeople feuding with farmers (Wright, 1988). While many believed the statue would be a 'great tourist drawcard', others argued that it was 'a travesty to glorify a man who shot three policemen' and that the statue 'transforms a vicious criminal into something "that looks like the Singing Cowboy"' (Wright, 1988, 3). One local resident was cited as saying 'for him to be glamorised in a romantic way on the basis of helping tourism is just rubbish. ... I believe this is the first time in the world that a government has paid to glorify a man who has shot policemen' (Bob Cummins cited in Wright, 1988, 3). The plaque attached to the statue acknowledges that Thunderbolt was a criminal but says nothing of victims. Despite the controversy, the statue remains today.

McCrossin's Mill Museum in Uralla (entry to the museum is $7 for an adult) dedicates its foyer to the Thunderbolt Paintings Series, which includes nine paintings by artist Phillip Pomroy depicting Thunderbolt's final day. The top floor of the museum also provides tourists with 'The Life and Legend of Thunderbolt' exhibit. The exhibit displays 'artefacts relating to Thunderbolt's daring escapades, including pistols and saddles he used and the table his body was displayed on at the Uralla Courthouse' (Uralla Visitor Information Centre, nd, 2). According to its website, 'McCrossin's Mill Museum houses the definitive collection of artefacts connected with the legendary Gentleman Bushranger, Fred Ward aka Captain Thunderbolt' (McCrossin's Mill Museum, 2010, para. 1).

**Figure 4.2:** Captain Thunderbolt statue, Uralla, NSW, Australia

Source: chris24 (nd), 'Statue of Australian Bushranger "Thunderbolt" at Uralla NSW'

Similar to Ned Kelly, tourists can embark on a 'Thunderbolt trailscape' by visiting a number of Thunderbolt-related locations across the New England region. For example, there is 'Thunderbolt's Rock', which is a natural granite outcrop that can be seen along the New England Highway heading out of Uralla towards Sydney (heavily graffitied now), which was used by Thunderbolt as a vantage point to attack mail coaches. Thunderbolt's cave, which was used as a hideout and is located minutes off the highway between Armidale and Guyra, allows those tourists travelling from Sydney to Brisbane (or vice versa) to visit another bushranger attraction.

In Tenterfield (population 6,628), two-and-half hours north of Uralla (and only half an hour from the Queensland/NSW border), Thunderbolt's Hideout is promoted. In addition to the hideout, tourists are also told about Thunderbolt's Tree and Thunderbolt's Leap (Tenterfield Shire Council, 2022). However, neither of these sites has easily identifiable directions. All these sites are situated within bushland and the hideout is the only site to have even minimum tourism infrastructure (signposting marking the site).

Unlike Ned Kelly, Thunderbolt is not heavily merchandised or commodified. Within Uralla, there is one café named after the bushranger: Thunderbolt Country Kitchen, and limited souvenirs are available, usually in the form of books about the bushranger. Despite this, merchandising opportunities have been taken advantage of by sellers both within and outside of the New

England region. A range of 'Thunderbolt' shirts, hats and facemasks can be bought online, many originating from American companies demonstrating the widespread awareness of the bushranger. A company, Thunderbolt Balms, has also been created in the New England region crafting and selling beard balms and oils. The labels for the balm and oil often feature a newspaper-type side-on profile sketch image of a bushranger (presumably Thunderbolt). Someone has even designed their own Captain Thunderbolt 'Pokémon card' (MyPokeCard, 2019).

## 'Trailscape' dark tourism

To create a successful 'trailscape' two key elements are required. First, a person or event needs to be turned into a worthy attraction that is capable of attracting tourists. Second, that person or theme needs to hold tourists' interest over a large geographical area, encompassing several communities or locations. Both Ned Kelly and Captain Thunderbolt have been commodified by several communities within a wider region, and tourists can undertake immersive travel across large stretches of rural and regional Australia.

Along the Ned Kelly Touring Route, there is an abundance of Ned Kelly themed souvenirs, and within Glenrowan, there is accommodation named after Kelly and numerous places to eat utilising the Kelly legend to theme their establishments. Each of the locations along the touring route has 'branded' their town through Ned Kelly marketing strategies, essentially taking 'control of the heritage' of Kelly and the narrative that has been told (Pearce et al, 2003, 74). While the narrative has been focused on Kelly as an Australian icon and 'larrikin', this narrative has become more challenged in recent years.

The ancestor of police victim, Michael Kennedy, believes that the 'lies' of the Kelly story are now being exposed. This may mean that conflict over which narrative is presented to tourists will be ongoing across all locations of the Kelly Touring Route. The victims' descendants and police organisational narrative would certainly be less attractive as a tourism strategy – moving away from the hero larrikin to a darker story of crime, violence and execution-style death.

The commercialisation of Thunderbolt also occurred to a lesser extent across the New England area. Numerous 'natural locations' have been branded as authentic sights of Thunderbolt's bushranging endeavours, and the feature of a permanent museum exhibit in Uralla and the Thunderbolt statue further 'marks' (and divides) the town. Unlike Ned Kelly, there are very few 'merchandising' options for Thunderbolt within the local area, and only one eating establishment is named after Thunderbolt.

Similarly, both bushrangers have undergone widespread (local, national and international) 'marketing' and 'packaging'. The Ned Kelly Touring Route has effectively created a 'do-it-yourself' tour of a large stretch of rural and regional NSW and Victorian towns. The use of the Ned Kelly helmet to 'brand' each location on a map effectively advertises the entire town as a significant heritage place associated with Australia's most famous bushranger. Importantly, the people of the local area play a large role in the 'marketing emphasis'; the staff at many of these themed locations and the shops selling souvenirs are well versed in the 'Kelly legend' and provide tourists with the highlights of Kelly's criminal career. As such, community members play a key role in telling stories. As already noted, the interpretation of the Kelly legend and, in particular the site of Stringybark Creek, has recently been reinterpreted highlighting the ever-evolving nature of tourism and the subjective nature of the information presented to tourists.

There is a similar 'do-it-yourself' atmosphere of the Thunderbolt region – interested tourists can take detours from the main highway to visit the relevant 'authentic' sites. McCrossin's Mill Museum similarly allows community members to provide tourists with the Thunderbolt story.

In both cases, there also seems to be community support (at least from some members) for the promotion of bushrangers as a tourist attraction. Tourism that focuses on one particular individual or event must have community support to thrive, and the narrative of the event needs to be 'in line with the community's vision for itself and for the kind of tourism it wants' (Pearce et al, 2003, 79–80). Given that Ned Kelly has often been romanticised and glorified as a national cultural icon, the choice to focus tourism marketing on the Kelly legend has benefited many communities. Yet, if there is a turning point in public opinion against Kelly, which there are indications of, it could mean that the towns along the Ned Kelly Touring Route will need to rebrand themselves. The tourism strategies surrounding Thunderbolt in Uralla have always been more controversial. The proposal and ultimate unveiling of the Thunderbolt statue divided the community, and it is perhaps because of this that the promotion of Thunderbolt (in terms of themed shops and souvenirs and so forth) has been downplayed and avoided. Despite this, both bushrangers offer local/national tourists an opportunity to explore sites that have significance to Australia's national heritage; and for international tourists, it offers light-hearted tourism of criminal activity and death reminiscent of other well-known international outlaws such as Jesse James.

## Conclusion

The admiration of bushrangers within popular culture in Australia is evident through continual tourism to rural and regional areas, as well as the wider national and international marketing of bushrangers like Ned Kelly and Frederick Ward. For example, both bushrangers appear on a range of merchandise at local, national and international levels.

The selling of bushranger imagery and narrative offers many rural and regional towns across Australia a unique marketing opportunity. Towns can sell the danger and mystery surrounding 'larrikin' outlaws, while simultaneously showcasing the beauty of the region. In many cases, there is no 'stain' around having bushrangers as an icon to celebrate and attract tourists, rather it is a badge of honour that the local surrounds housed, and even protected, these criminals.

However, as seen throughout this chapter, there are varying (and evolving) perspectives within the local communities on how bushrangers should be portrayed (and essentially marketed). Concerning Ned Kelly, there are ongoing debates about which narrative should be paramount – the 'hero' bushranger, or the 'criminal' bushranger with an emphasis on his victims. Such changing sentiments towards Ned Kelly have led to altered narratives in official tourism documents and sites, as well as the erection of memorials to the victims, and their advertisement as part of the Ned Kelly trailscape. The debate has even transcended the local community – being questioned at a national level whenever Ned Kelly should be used as a cultural icon. Similarly, in Uralla, the community was divided over the erection of the Captain Thunderbolt statue, which resulted in neighbours refusing to speak to each other. As such, 'dark tourism' can create long-term animosity within a community. In the case of both Ned Kelly and Captain Thunderbolt, the tourism aspect 'won out' with sites, shops and hospitality establishments relying on the bushranger tourism industry.

Ned Kelly offers an immediately recognisable visual representation – his helmet and armour have become famous national icons. This, in turn, offers numerous souvenir and merchandising opportunities that can be 'authentically' sold at each location along the touring route (and more broadly across Australia). In the case of Thunderbolt, where there has been more 'shame' attached to his criminal acts, there is still honour surrounding the name Thunderbolt (with natural locations being renamed in his honour – and a statue and museum being established in Uralla) – yet his legacy is not promoted, packaged and sold to the same extent as Kelly. This could also be related to the broader knowledge of both bushrangers across Australia. As already noted, most Australians can easily identify Ned Kelly, whereas only a small proportion are aware of Thunderbolt.

While there is more familiarity with Ned Kelly, particularly within Australia, most 'locals' within a community tend to know their local bushranger history and heritage. This local folklore is then passed along to newcomers and tourists alike as a selling point of the town and its long historical heritage. There is also some international recognition and appeal of Ned Kelly, with several movies made about the outlaw and famous actors such as Mick Jagger and Heath Ledger representing Ned Kelly. As noted, there are also international examples of famous bushrangers, particularly in the United States, which make bushranger sites relatable and attractive to international tourists. In the case of both Thunderbolt and Kelly, there are multiple (authentic and replicated) physical reminders of their lives. There are (in some cases recreated) homes, schools, graveyards, natural sites such as camp sites, museums and interpretive sites that are well sign-posted and relatively easily accessible for tourists.

Within themescapes or trailscapes, the landscape has a vital role to play in facilitating and selling a story. For example, in the case of bushrangers, without the bushland surrounding these towns, or the authentic physical reminders of the bushranger activities being maintained, tourism activity would likely stop. The geographical isolation and beauty of the bush landscape help to 'sell' the story to tourists, and the theming of a route throughout this more isolated and rugged landscape provides a tourism strategy for a wide area. Similarly, the natural rock outcrops and caves featured in the Thunderbolt themescape, or the numerous sites along the Kelly Trail, provide tourists with a comprehensive and fully immersive tourism experience. Such locations are advertised as 'time capsules' of the bushranger story, with the natural landscape changing very little, and many of the towns also remaining 'frozen in time' with many of the original structures still present for tourists to visit.

As Fagence (2017, 455) has noted, the 'geographical filter' of visiting bushranger sites provides tourists with a 'sense of space' as well as the 'significance of place' and distance. However, without the 'texts' or tangible artefacts, documents, and so forth, the place loses focus, and perhaps this is another reason why tourism to Thunderbolt is substantially less, simply because there is very limited interpretation at the sites and almost non-existent tourism infrastructure. The narrative that guides tourists' 'gaze' as to what is important at a site helps to illuminate the distinctiveness of the geographical location. Having authentic exhibits within these locations deepens tourists' appreciation of the events that happened. While there are exhibits available in the McCrossin's Mill Museum in Uralla, it is missing at other Thunderbolt locations. In contrast, there is some form of signposting and storytelling at all of the Ned Kelly tourism locations.

Urban tourist centres also have a vested interest in promoting the bushranger narrative. Themes of rural vastness and complexity should dominate the storyline of any bushranger tourism destination, yet, in contrast, the geographical location is not explored in any significant detail in urban tourist destinations relating to Ned Kelly or Thunderbolt. Melbourne, in particular, promotes the Ned Kelly narrative, and while acknowledging the Kelly story of rural areas, focuses more on the urban elements or fails to convey the rural landscape and geographical intricacies of policing bushrangers in isolated and rugged terrains.

The rural and regional narratives surrounding both Ned Kelly and Captain Thunderbolt centre upon both men's skills as 'bushmen' and their ability to evade police capture by utilising and understanding the bush and its surrounds better than the urbanising influence of law enforcement. The bush plays a key role in the bushranger mythscape – without the untamed, isolated and dangerous landscape, there is nowhere for outlaws to live or hide. As many of these sites have remained undeveloped, they provide an authentic tourist destination that enables tourists to 'experience' rural landscape and the thrill of danger and crime within the knowledge that civilisation is just a short car trip away.

The 'darkness' of the bushranger plays a role in the attractiveness of tourism and wider popular culture awareness and attention. As Tranter and Donoghue (2008, 384) hypothesise, 'Why is Ned Kelly so well known, when most other bushrangers were named by so few Australians? … they were less violent than Kelly. They did not kill and, in Thunderbolt's case, did not even shoot anybody'. The lack of violence makes such tourism less 'dark' but also less 'attractive'. As such, tourism for less violent bushrangers is more localised, while Ned Kelly attracts national (and even international) attention. For Tanter and Donoghue (2008, 386), the instantaneously recognisable helmet and armour also play a large role in the ongoing fascination with Ned Kelly – with the armour becoming 'an Australian symbol for bravery and resistance against injustice and oppression'.

Promoting Ned Kelly in such a manner assists in the promotion of more light-hearted and 'gimmicky' tourist sites and merchandise. The comparison of the Glenrowan animatronic theatrical retelling to a blend of Hammer Horror (or gothic horror) and Disneyland indicates the 'lightness' to which Kelly's criminal antics have been approached and the ease with which such attractions promote family-friendly tourism destinations. This seems contradictory – that the more violent of the two bushrangers has been rethemed into a light Disney, albeit 'gothic', type attraction, while the less violent bushranger has had seemingly more local controversy and less allure.

Many ethical questions arise regarding the 'selling' of violence and the glorification of bushrangers to tourists and particularly to young children. As noted, communities and descendants are divided over whether bushrangers should be seen as heroes or villains, and this debate continues more broadly on local and national platforms. While the ethics of some tourism 'gimmicks' have been questioned, such as with the statue of Captain Thunderbolt in Uralla, others have received little attention. For example, the ability for children to dress up in Ned Kelly's replica armour at the Old Melbourne Gaol raises two important ethical questions that are neglected by the site and the wider literature.

The first relates to offering children the ability to dress up as outlaws, thus further glorifying violent criminals. However, in our society, children are often allowed (and even encouraged) to dress up as criminals or 'monsters' from popular culture, including serial killers (both fictional and non-fictional), particularly for Halloween, but also for Book Week parades at schools. As such, the question of dressing up as Ned Kelly for a quick photo seems within the realm of ethical sensibilities. However, the second ethical consideration in this case relates to the *location* of this dress-up option. Would parents happily encourage their child to dress in the armour and pose in the location if they were aware that this was the drainage spot where the hanged people's last excreted bodily fluids were washed away? This would likely raise more uneasiness about this activity, compared to putting on the armour itself.

What is clear is that the narrative given to bushrangers in these locations (for example, are they portrayed as local heroes or is the focus on the crimes they engaged in) and the situational rural context of the locations themselves is critical to the selling of bushrangers as a source for (dark) tourism sites. According to Fagence (2016, 13), folk heroes are important for local economies and, as such, their legend is propagated and focused upon to maintain tourist activities. In the case of Kelly and Thunderbolt, a range of 'stories' have been presented, yet the dominating theme across both locations has been the focus on the 'criminal' activities, and in Kelly's case, sites of death, turning these locations into dark tourism sites.

PART II

# Tourism Sites of Recent Controversy

5

# Carceral Tourism

Carceral tourism incorporates criminal justice sites that have been characterised by inflicting pain on past inmates through the administration of justice and includes prison, police and courthouse museums (Piché and Walby, 2018). While Welch (2013, 479) argues that prison museums invert the 'Disney' experience, and essentially become the 'antithesis of "the happiest place on earth"' (Williams, 2007, 99), many, if not all carceral tourism sites in Australia sit on Stone's (2006) lighter side of the dark tourism spectrum, despite the often 'heavy' subject material that these sites deal with. Most of these criminal justice tourist sites can be categorised as dark dungeons', providing tourists with 'a combination of entertainment and education as a main merchandise focus' at the original site of justice (Stone, 2006, 154).

Visitors to these sites are often encouraged to participate in the criminal justice process – be arrested, sentenced, locked up and maybe even threatened with execution. At night, tourists are invited to participate in ghost tours and are told even more gruesome and confronting stories of justice being enacted (or evaded). Stone suggests that dark dungeons should occupy the centre ground of the spectrum because such sites are authentically located (at a decommissioned prison or courthouse, for example) and the material being dealt with is necessarily 'dark' and confronting. The reality is that in many cases the 'entertainment' wins out, and these sites are more 'light' than 'dark', with high levels of tourism infrastructure catering to a range of audiences, often including schoolchildren.

This may be because carceral tourism within Australia is a mix of dark dungeons and dark fun factories. There is a commercial ethic at these sites, and while they present real 'macabre' events, the entertainment focus can overshadow the darker messages being presented.

## Penal sites

Penal (prison) museums and tourism sites allow the bystander to 'gaze at the spectacle of pain and suffering' while keeping 'spectators at a safe social distance from the realities of cruelty (for example, torture)' (Welch, 2013, 480). Reasons why tourists visit penal museums are varied: 'some visit out of curiosity, others visit to remember loved ones' (Walby and Piché, 2011, 452) or 'to retrace history, to search for ghosts, and to view otherwise prohibited places' (Brown, 2009, 90). Further:

> The politics of the penal gaze here are inevitably embedded in spectacle and thrill-seeking. However, these tours also serve as experiences that, once completed, can be claimed as exemplary and authentic, grounded in institutional and historical experience. (Brown, 2009, 91)

Regardless of the motivation, the success of prison tourism hinges on providing tourists with 'an opportunity to gain a glimpse of life "inside"' (Wilson et al, 2017, 3). Within Australia, it has been argued that there is limited opportunity for dark tourism sites (Wilson 2008a), and that 'former gaols dominate the scene of dark tourism in Australia' (Shehata et al, 2018, 7).

There has been a long history of penal tourism, both within Australia and worldwide. According to Wilson (2011a), penal tourists played a key role in the reforms in the design of prisons in Britain in the first half of the nineteenth century, and architects also used prison tours to sell facility designs and best practices to prison authorities (Piché and Walby, 2010). Yet, as Wilson et al (2017, 4) argue, carceral, and in particular 'prison tourism is big business, and has the potential to contribute to many facets of a community's cultural understanding, children's education, economic benefit, and even its international profile'. As part of the allure offered to tourists, devices for maltreatment, torture and execution are often displayed alongside 'scientific' or 'philosophical' explanations for the use of such equipment throughout history (Welch, 2013). In more recent times, prison tours (of functional corrective centres) have been used 'as a deterrence mechanism to "scare straight" so-called at risk youth' (Piché and Walby, 2010, 570).

Historical penal museums can offer tourists the opportunity to reflect on past (and current) social, political and cultural norms and shifts in values and approaches to administering 'justice'. Yet, the balance between educating and entertaining visitors is often fraught in penal museums, with many sites choosing to 'entertain, titillate, amuse or "frighten" the average member of the public, with the added imperative in many cases to preserve the "establishment" narrative pertaining to the former institution' (Wilson et al, 2017, 5). For Wilson et al and others, there is a danger of tourists leaving such sites with a:

> misconception that the narratives presented are a fair and complete representation of the institution and its former residents, when those narratives are in fact narrowly conceived and deeply contested, and

what was presented as 'authentic' was very much 'staged' ... to create palatable, sanitized forms of history and memory. (Wilson et al, 2017, 5)

The danger is that visitors leave with reinforced notions of the state form of 'justice' and a lack of understanding, or empathy, for those imprisoned. In many cases, it is only the 'celebrity' prisoner voices that are 'heard' – and this may be an intentional selling tool for many rural sites. While these celebrity voices may not reflect the experiences of all prisoners, they serve a purpose in attracting attention to the site itself.

Penal museums are, on the surface, difficult places to 'sell' – they provide evidence of the pain and suffering of individuals at the hands of other people and the government. Yet, most penal museums utilise a range of tactics to 'lighten' the mood and present a palatable experience for tourists. Guided tours, one of the key experiences for many tourist sites, are often couched in 'comfortably humorous terms, focusing chiefly on the supposedly "lighter" side of prison life such as escape attempts gone wrong, stories of eccentric (that is, harmlessly amusing) characters among the inmates and so on' (Wilson, 2011a, 567). Anecdotes of more 'serious' material such as suicides and guard and inmate brutality are rarely dealt with, or if they are, in fleeting moments that downplay such behaviour as 'uncommon tragedies' (Wilson, 2011a, 567).

For the most part, visitors to decommissioned penal museums invert 'the dark tourism norm in that the attitudes of mainstream visitors to the sites by and large *endorse* the suffering of the victims' (Wilson, 2008b, 333). The inmates are 'othered', and while there is 'partial empathy' for inmates (Wilson, 2008b, 333), there is little questioning of the 'justice' administered by the criminal justice system, or that offenders, through transgressing social norms 'deserve' such punishment. In other cases, tourists are invited to look at, or enter, small cells or view torture devices where the museum denounces these types of violence inflicted on inmates before the 'so-called penal reform projects of the mid-20th century' (Walby and Piché, 2011, 459).

The result is that penal museums in Australia provide a 'safe' dark tourism destination where visitors can be exposed to pain, suffering and macabre concepts and events while remaining entertained and secure in the belief that only 'bad' people (mostly) suffered here and that the suffering of inmates is mainly historical with many brutalities 'removed' from modern prisons – in reality, many brutal practices remain worldwide, including small cells such as the 'hole' and physical restraints for inmates which has resulted in deaths (Walby and Piché, 2011, 460).

The 'othering' of inmates has enabled some of these sites to be converted from a place of misery and uncomfortable histories into cultural venues such as sites for public theatre and a culinary school (Sandhurst Gaol), a school of art (Long Bay Correctional Centre), and 'mixed-use development including hostel, commercial and residential facilities' (H.M. Pentridge) (Shehata et al, 2018, 2), while others that have been converted into a museum are still advertised as sites for formal occasions, such as weddings (Old Melbourne Gaol, Old Dubbo Gaol).

Because museums are housed in decommissioned prisons, they are automatically judged as authentic by members of the public, and where executions have occurred, these sites are treated as 'hallowed ground' (Welch, 2013, 483). While numerous research within Australia, and overseas, has focused on the authenticity of the material presented in these dark tourism sites, there has been little to no research on how rural and regional penal museums differ from urban sites or how the geographical differences in penal practices are conveyed to visitors.

Most nineteenth-century prisons were designed and constructed in an imposing, neo-gothic manner. Wilson argues:

> the prison was designed specifically to exude an air of towering potency over the landscape, and so, for the edification of those free citizens observing from outside, to express, in the most confrontational possible manner, a sense of dreadful consequences, and of the righteous might that wreaked those consequences. (Wilson, 2011a, 565)

With the closure of these prisons, the aesthetic appearance took on a 'romantic' sensibility, providing significant motivation for tourism (Wilson, 2011a, 565). As outlined in Chapter 1, most prisons were built outside of cities within a more regional environment. Yet, as cities have expanded, many of these sites became absorbed by the 'urban' and this has helped shape its transformation as a tourist destination site. For example, 'urban' decommissioned prisons become surrounded by 'urban' sites, noises and routines to the extent that some of these sites become mini communities, such as the Pentridge Prison in Victoria, where tourism plays a small role in its economic rebirth. Shehata et al (2018, 7) found that 'for many, the proximity to CBD [central business district] appear to overcome dark memories of the site'. For those decommissioned prisons in regional and rural areas, the architecture can continue to stand in stark contrast to the growing community, yet as we saw with Port Arthur, this also lends itself to providing a romantic visual attraction.

## Ararat's Old Gaol and J Ward

It has been claimed that 'Ararat offers the best example in the country of dark tourism. ... We are the only town with two lunatic asylums [Aradale Asylum and J Ward in Ararat's Old Gaol]' (Edge Insights, nd, 22). Despite past statistics indicating that most tourism in rural/regional areas is domestic, Ararat claims that it has an international and national market of 'ghost hunters, paranormal investigators, thrill seekers, history buffs and many others' who visit the town for these dark tourist activities (Edge Insights, nd, 28). The site attracts 11,000 people a year despite being two-and-half hours from Melbourne (and the town itself only has a population of 11,965 people).

The Ararat Old Gaol opened in 1861 to provide incarceration facilities following the district's gold rush. When it opened, it housed 21 prisoners, increasing to 40 in 1864 (Only Melbourne, nd). On 15 August 1870, the first execution was conducted at the gaol, with two more to follow in 1884 and 1886. It was closed in 1886 as a gaol but reopened just a year later as a temporary hospital for the 'criminally insane', which was known as J Ward. Here, 'the most depraved and dangerous men in Victoria were housed in horrific conditions under the highest security' (Ararat Asylums, nda, para. 1). Although designed as temporary lodging, J Ward was not closed until 1991. Just two years later, the Friends of J Ward volunteer organisation formed, and tours of the site began. Given the site has only recently closed as an active institution, it makes J Ward 'darker' because many former employees and patients live locally, and many of the community members have 'living' stories of the site.

Ararat is a rural city surrounded by many natural attractions and national parks. It is situated within the Grampians tourism region on the Western Highway with direct road and rail linkages to Melbourne, Ballarat and Adelaide. In the mid-2010s, Ararat commissioned a Visitor Economy Strategy for 2018–2021. At the time, tourism to the town was on the decline despite two million visitors entering the Grampians region each year, and the economic strategy document was important because 'tourism is critical to the economy of the region' (Edge Insights, nd, 8). The Strategy identified J Ward and Aradale as major attractions and foresaw these sites as 'offering access to a niche market' with 'the potential to be an economic driver for the region' (Edge Insights, nd, 22).

As such, it was recommended that Ararat actively embrace being labelled a 'dark tourism' destination to increase tourism to the town. The strategy recommended that J Ward should be transitioned from a volunteer model to a corporate model and for increased tourism infrastructure to be added, including boutique accommodation on the site. Despite these recommendations, it is clear that, while J Ward remains a prominent tourism destination point for those visiting the Ararat region, the town did not implement the two main recommendations. The webpage advertises that daily tours are still conducted by approximately 70 volunteers (Friends of J Ward, 2017a). Despite the lack of corporatisation of the site, visitors rate the J Ward Museum highly, with Tripadvisor ranking it 4.5 out of 5 (based on 271 reviews), the number one tourist destination in Ararat and the site winning the Tripadvisor Travellers' Choice award for 2021 and 2023 (these awards are provided to accommodations, attractions and restaurants that consistently receive positive reviews and ranked within the top 10 per cent of attractions on Tripadvisor). Indeed, some comments specifically highlight the role of volunteers and how those volunteers enriched their experience. The Aradale Lunatic Asylum and J Ward Lunatic Asylum Ghost Tour also ranked 4.5 out of 5, however, both were ranked lower on the places to visit in Ararat.

The website encourages tourists to 'hear stories of its time as a gaol with murderers and thieves and later as the home for Victoria's criminally insane' (Friends of J Ward, 2017b, para. 1). Several different types of tours are offered – some by volunteers, and others by external organisations. For example, Ararat Asylums offer daytime tours, utilising Friends of J Ward volunteers to host the tours. The daytime guided tours, while being advertised as 'no two being the same' (because each guide has their own 'favourite true stories' to recount), in general, promise tourists:

> You will see the place where murderers breathed their last. You will see where their bodies are buried in unconsecrated [sic] ground in unmarked graves. You might hear the story of the notorious Garry Webb or of Bill Wallace who lived here for over sixty years and died in his 108th year. You might hear how the Chapel was built due to the efforts of William Watson Carr. You might hear of escapes or suicides or of twelve year olds incarcerated here. (Ararat Asylums, nda, para. 4)

For the day tours, tourists are advised that a gift shop and light refreshments are available for purchase. Interestingly, visitors are also forewarned that tour guides do not have access to 'records or have knowledge of former patients or medical procedures' (Ararat Asylums, ndb, para. 4). The site is seen by the volunteers as 'stand[ing] as a grateful reminder of how we've evolved from the brutality of past centuries' (Alex Beveridge cited in Hudson, 2017, para. 18).

The three night-time tour options are run by an external company, Lantern Ghost Tours. The first is a two-hour night-time ghost tour, which is advertised as follows:

> Explore the museum followed by a ghost tour of the home of notorious criminals including Chopper Read, Gary Webb and William Wallace. Explore the governor's bathroom, hangman's gallows, original kitchen, showerblock, grave sites, West Wing, J Ward Block, exercise yards and grounds in search of the souls that still linger. Learn of governors whose spirits are still lurking, prisoners buried in the grounds and the many souls trapped within these walls. (Ararat Asylums, ndc, para. 2)

For those who want to investigate the paranormal, up to ten tourists can pay to stay for an extra hour from the ghost tour to investigate hot spots (the second type of tour available) and use 'paranormal investigation equipment to make contact with the otherside' (Ararat Asylums, ndd, para. 4). The third night–time tour option offers an extensive 11-hour overnight ghost tour, starting at 8 pm and ending at 7 am. It is marketed as an 'all–inclusive' tour of food, sleep and, of course, paranormal activity:

> Join us on the ultimate overnight paranormal investigation. Visit Australia's home to the criminally insane, J Ward Lunatic Asylum. Your night starts with a 2 hour ghost tour of the home of notorious criminals including Chopper Read, Gary Webb and William Wallace. Enjoy a casual supper of pizza and soft drink in the original dining room and then follow our expert paranormal investigation team as they explore the governor's bathroom, hangman's gallows, original kitchen, showerblock, grave sites, West Wing, J Ward Block, exercise yards and grounds in search of the souls that still linger. End the night sleeping in the original, unrenovated gaol cells. (Lantern Ghost Tours, 2021, para. 1)

Accommodation 'is in original, unrenovated cells' and tourists are advised that they need to bring their own bedding (Lantern Ghost Tours, 2021). The tours are clearly aimed at titillating and entertaining tourists. Through highlighting famous prisoners (Chopper Read), and telling tourists that 'souls still linger', this site is very much a combination of a dark dungeon and a dark fun factory. Indeed, the lightness of the site is enhanced by being available for hire for weddings, plays, school socials and other functions, like other rural (and urban) dark tourism sites.

## Old Beechworth Gaol

Historic country gaols, like urban gaols, were designed for impact. That is, they were usually located in easily visible places outside the town boundary so that people in town were continually reminded of 'what happened to criminals, and to deter them from crime' (Edmonds, 2019, 55). Beechworth Gaol was constructed on a piece of land gently sloping away from the town so that it could be laid out symmetrically with generous yards while remaining 'visible' to the town.

Beechworth was 'the centre of one of Victoria's richest goldfields in 1853' (Heritage Victoria, 2014, 11). In 1856, Beechworth Gaol began as part of a government camp (Collins, 2021a) as a temporary stockade. However, the geographical location of Beechworth meant that prisoners serving long sentences for serious offences were often prevented from travelling to

Melbourne (Heritage Victoria, 2014), aiding in the decision to establish a permanent prison at this site. The gaol was based on the 'Panopticon' design, however conditions were primitive – for example, it was not until 1993 that the prison was provided with flushing toilets or running water (Heritage Victoria, 2014, 3). While construction of the gaol was still incomplete, it opened officially on 3 July 1860, providing single cells for 36 prisoners, which increased to 75 cells at the time of the gaol's completion in 1864. From the time it was opened until 1895, the prison housed both male and female prisoners who were required to work 'for the benefit of the town' (Collins, 2021a, para. 9). For men, this meant crushing rock for the construction of government buildings or road base, and women were required to wash and sew for government officials. Gallows were added to the design in 1864, with eight men being executed between 1865 and 1881 and buried in unmarked graves on site.

According to Edmonds (2019), the first prisoners to be housed in Beechworth Gaol were Chinese men from the goldfields. Originally housed in the Sandhurst stockade, the Chinese prisoners were kept together in one cell away from the 'European' prisoners due to the belief that they were of a lower social order (Edmonds, 2019). This indiscriminate grouping of all Chinese men into one cell led to 'friction' between the men, resulting in further discrimination by criminal justice authorities. As such, when Beechworth was opened with individual cells, the Chinese prisoners were detained to allow authorities to 'maintain better order and discipline among them' (Public Records Office of Victoria cited in Edmonds, 2019, 53). Beechworth Gaol also housed Ned Kelly for six months between 1870 and 1871 for assault, as well as in 1880 for his committal hearing for murder, attracting further tourism to the area.

Due to the extension of the railway into Beechworth in 1876, there were declining numbers of prisoners being sent to Beechworth Gaol (instead, prisoners were sent to Pentridge in Melbourne), resulting in the repurposing of the gaol several times. The prison was decommissioned and closed on 20 December 2004. The following year, the prison was sold by Corrections Victoria, a government instrumentality, to a private family, and six years later opened as a heritage-listed tourist site in 2011. Heritage Victoria (2014) saw the site as historically significant because of its association with the major prison construction program that occurred in Victoria around 1859–64. It was also viewed as significant because it:

> demonstrates the conditions under which prisoners were kept in the mid-nineteenth century, when prisoners were subjected to solitary confinement in single cells without adequate lighting, heating, sewerage or running water. … It is of historical significance for its associations with the bushranger Ned Kelly. (Heritage Victoria, 2014, 4)

The archaeological design of the building, as well as the burial sites of the eight men executed, were also highlighted as being archaeologically significant.

However, despite this focus from Heritage Victoria, the site has been referred to as 'neglected', and in 2016, the Australian Centre for Rural Entrepreneurship (ACRE) led a $2.5 million 'consortium of 20 local families and philanthropic organisations in a local buy-back' of the prison to create 'a vibrant precinct for the social, cultural and economic benefit of the region and a centre of excellence to inspire rural communities across Australia' (Old Beechworth Gaol, nd, para. 1). The gaol now houses the ACRE not-for-profit organisation, as well as Breakout, which offers office space for small entrepreneurs in the region that would like to collaborate. According to ACRE, 'the Old Beechworth Gaol has an ambitious growth agenda, centred around building a world-class, cultural tourism asset and visitor experience' (Scully, 2021, para. 3). Approximately 45,000 visitors participate in gaol tours and the site hosts 15 community festivals and other cultural events.

Those who participate in the tour are labelled as 'inmates' and tours last approximately one hour (Tripadvisor, 2021a). There is also a Paranormal Prospectors Australia tour available to adults over the age of 18, where tourists explore paranormal activity in the gaol for three hours. Tripadvisor ranks the prison museum a 4 out of 5 (based on 429 reviews), slightly behind the other sites discussed in this chapter (although it has had more reviews than some of the sites with higher ratings), as well as the fourth best thing to do in Beechworth. In addition to the tour, it has a café, which doubles as a gift shop to sell T-shirts and beanies.

The Old Beechworth Gaol is undeniably a site of 'darkness'; it saw punishment, execution and other causes of death. In addition to the eight men who were formally executed at the site, there have been at least 40 additional deaths in custody – many from suicide. Local historians and researchers, including Nanette Collins, provide a comprehensive website detailing the prison history of the site, as well as the deaths that occurred on site. The reasons for death include strychnine powder, diarrhoea, suffocation, asphyxiation, hanging, heat apoplexy, old age, dysentery and 'visitation from god' (otherwise known as dying from natural causes) (Collins, 2021b). Despite this, the site still offers tours and souvenirs, allowing lighter, family-orientated entertainment, as well as more 'dark', adult-orientated night tours.

## Old Dubbo Gaol

Dubbo's history of law enforcement began in 1846 when it held its first Court of Petty Sessions. At this time the town was not recognised as a village – this did not occur until November 1849, and before 1846, all criminal matters were dealt with in Wellington. The Old Dubbo Gaol (see Figure 5.1) was

**Figure 5.1:** Entrance to the Old Dubbo Gaol, NSW, Australia

Source: Quixley, J. (nd), 'At the Old Dubbo Gaol'

one of the first government buildings in the area and began as a courthouse lockup in 1847 before becoming a gaol in 1859 (although there were several stages to building the gaol, which continued through to 1871). The first execution to be carried out at Dubbo Gaol took place on 29 May 1877. There were a further seven executions between 1880 and 1904. In 1914 the gaol was demoted to a police gaol, which meant that hanging could no longer occur on site, and only prisoners serving short sentences could be housed there (Dubbo Regional Council, 2021a). On 31 August 1966, the gaol was officially closed and the remaining prisoners were transferred to Bathurst or Parramatta.

Just eight years later, the site was reopened as a tourist attraction after being repaired by the Old Dubbo Gaol Restoration Committee. Unlike other rural sites that have used the lure of violent and monstrous prisoners and crimes to entice visitors, the Old Dubbo Gaol website takes a slightly different approach:

> Where history comes to life, this 19th century prison explores the brutality of prison life, the daring escapes and heritage of the site delivered in a mix of theatrical performances and modern immersive displays. ... Explore the gaol in your own time or catch (literally!) our characters in costume as they dare to outsmart the Gaol Warder. For some tall tales and performances by our animated wardens and prisoners

check out our tours and performance schedule for entertaining stories on prison life and their attempts to breakout. Good behaviour is rewarded so purchase your tickets in advance online and receive a 10% discount. (Dubbo Regional Council, 2021b, paras. 1, 2)

Further, when sharing the history of the site, the website advises tourists that the prisoners housed in Old Dubbo Gaol were victims of the growth in the town and region, and committed crimes because of 'greed, hunger and poverty' (Dubbo Regional Council, 2021a, para. 2). In contrast to most of the sites covered in this chapter, parts of the site (and activities) at the Old Dubbo Gaol offer fun, light-hearted entertainment that is family-friendly. At the same time, a 'darker' narrative is also offered to visitors with narratives of hangings and hardships encountered by inmates. As such, the site caters to a range of tourists and recognises the differing motivations for travel, as well as the need to provide activities for children. Further, the site promotes the idea that the prisoners housed at the site were not necessarily 'evil' or 'violent' but victims of circumstance and life.

The prisoners are to be pitied rather than feared, and the website advertises that tourists will be able to 'get a sense' of the loneliness felt by even the 'most hardened' criminals in the male cell division (Dubbo Regional Council, 2021c, 'male cell division'). The site also naturally deals with the executions that took place within the grounds. Tourists are invited to visit the Gallows Gallery to 'See prisoner's [sic] whose crimes sent them to the gallows. This extensive collection of a hangman's kit overlooking the gallery of condemned men may cause the hairs on your neck to stand' (Dubbo Regional Council, 2021c, 'the gallows gallery').

While tourists can walk around the site unguided, there are also different types of tours and performances available. The 'Escape Talk and Characters in Costume' theatrical performance sees 'characters' re-enacting breakout attempts. This performance is included in the normal price of admission, so visitors just need to make sure they arrive at the site at the relevant time to catch this entertainment. After-dark tours are also available during school holidays. Children can also participate in a 'Super Sleuth' challenge to keep them occupied throughout their visit.

The Twilight tour starts at 6.15 pm and is advertised as 'an all ages theatrical tour fun for the family. Be inducted as the newest inmates and hope your behaviour sees you released by nightfall' (Dubbo Regional Council, 2021c, 'Twilight tour'). A few hours later, the Beyond the Grave tour starts at 8.45 pm for adults only and is marketed to 'those who enjoy the spine tingling tales of death, punishment and darkness. Explore the cells by torchlight and sense the suffering of what still may linger behind our walls' (Dubbo Regional Council, 2021c, 'Beyond the Grave'). Tours can also be organised for school groups to allow children to 'escape the classroom' or to accommodate private groups of 15 or more people (Dubbo Regional Council, 2021d).

The website has been designed to tie all aspects of tourism into the language of crime and punishment. For example, the gift shop is advertised as follows:

> Raid our museum shop for souvenir and provisions. Grab a souvenir distinctly unique, from produce and merchandise with regional flair to convict keepsakes perfect for the Warder-in-training or sneaky escapee! (Dubbo Regional Council, 2021b, para. 3)

Similarly, the advertisement of the site for venue hire is promoted in the following way:

> Events in The Yard: The Yard, in any Gaol was a place of exercise, a place of work, a place of muster, and for precious few moments each week a place of recreation for prisoners. The Yard offers an event venue that easily accommodates weddings, music, markets and more. (Dubbo Regional Council, 2021a, para. 1)

The gaol is open 364 days a year and has a clear dark fun factory appeal – the site tours and marketing promote immersive family-friendly entertainment. The site is advertised by several tourist sites, including Tripadvisor (2021b, para. 1) as 'a state heritage listed must see'.

Interestingly, the advertisement of the Old Dubbo Gaol on other external websites such as Tripadvisor, reverts to a more traditional 'dark' sell:

> This regional gaol is representative of powerful, surprising and dark moments in Australian prison history. We house important collections such as the hangman's kit and gallows, and unforgettable experiences such as the dark cells and the bird's-eye view from the watchtower. (Tripadvisor, 2021b, para. 1)

Emphasising the hangman's kit and gallows and the 'darkness' of the site conveys a different message than that of the Old Dubbo Gaol website, and it may be a way of attracting 'dark' tourists that have visited other dark tourism sites to Dubbo.

## Trial Bay Gaol

The ruins of Trial Bay Gaol are located in the Arakoon National Park in NSW near South West Rocks (the ruins can be seen from this township), and just over an hour south of Coffs Harbour. The ruins include:

conserved penal amenities such as a kitchen and archaeological vestiges of a hospital; two double-storey cell blocks and a silent cell block; and two open prison cells wherein tourists listen to looped recordings of re-enactments of inmate conversations. (Barnes and McIntyre, 2017, 61)

While Trial Bay Gaol does not attract the same level of tourists as Port Arthur or Old Dubbo Gaol, it has still recorded around 36,500 visitors a year since 2006; of these visitors, 32 per cent are from the surrounding region, 60 per cent from elsewhere in Australia, and 8 per cent are international (Barnes and McIntyre, 2017, 60). Visitors rated Trial Bay Gaol 4.5 out of 5 on Tripadvisor (from 660 reviews), on par with other more famous rural/ regional sites such as PAHS and Old Dubbo Gaol, demonstrating that smaller dark tourism sites are just as appealing,

Construction of the gaol began in the 1870s, during a time of penal reform across the British Empire. The gaol was designed with the reformist intention of housing long-term prisoners nearing the end of their sentence who could be put to work constructing a 1,500-metre breakwater to make the bay safe for ships sailing between Sydney and Brisbane. The remoteness of Trial Bay from urban centres served as a natural deterrent for prisoners considering escape (Davies, 2004), and the cleaning of natural vegetation enabled further surveillance and created an atmosphere of foreboding (Barnes and McIntyre, 2017). The prison structure was imposing: constructed of granite (which required quarrying machinery to be brought in from Sydney), the prison had a central hall with a single two-storey cell block, which was enclosed by a high stone wall with four watchtowers (Davies, 2004), thus enabling simultaneous surveillance of prisoners within cell wings. The gaol also housed five silent secondary punishment cells to cater to the separate treatment system. While these cells were used infrequently at Trial Bay, they were described as 'gloomy crypts into which light never penetrated' (NPWS cited in Barnes and McIntyre, 2017, 62).

In 1903, the gaol was closed after it became clear that the breakwater would not succeed due to continual storm damage. In 1915, the empty buildings were repurposed as an internment camp for Germans living in Australia who were suspected of sympathising with German sentiments. Despite being imprisoned, tourists today are disinclined to view their time with much sympathy:

It wasn't exactly a hard labour camp, more like a holiday camp, despite wartime restrictions. For a start, because the South West Rocks site was considered so remote, the internees had a fairly idyllic, even tedious, lifestyle, being able to freely wander about outside the sombre prison walls. After all, where would they go? In every direction, there was bush or the ocean. (Scanlon, 2019, para. 31)

The German internees built a 20-foot granite monument above the Trial Bay Gaol to commemorate their comrades who had died at the prison and elsewhere during the war. The monument was a source of community conflict and, in 1919, it was destroyed by 'some person, or persons unknown' (The Sun, 1919, 1). This led to 'informal, dark-minded tourism' (Barnes and McIntyre, 2017, 64) arising with 'hundreds of people' visiting the ruined monument within just the first week (The Urana Independent and Clear Hills Standard, 1919, 1).

Broader tourism to the site, much like Port Arthur, occurred spontaneously and naturally. In 1886 one local person reported weekend boat excursionists arriving to tour the site, reporting that he had 'never [seen] such a crowd here before' (Ennis cited in Barnes and McIntyre, 2017, 65). Following its closure, the prison remained a tourist destination, and in 1906, the site was included in the Government Tourist Bureau's first organised group tour to the Macleay. This group tour reportedly 'delighted' the visitors (Clarence and Richmond Examiner, 1906, 4). In the 1930s, concern over the preservation of the ruins was raised by three influential travel writers after vandalism occurred at the gaol and continued over the following decade as the site continued to fall into decay (Barnes and McIntyre, 2017).

The current level of tourism infrastructure at Trial Bay Gaol is modest compared to other sites such as Port Arthur or the Old Dubbo Gaol. It offers a visitor information desk, an orientation area, a small museum and three guided tours (as well as camping grounds for overnight tourists). The first of the three guided tours, 'Walk on the dark side: Sunset tour', advertises that tourists will experience 'the darker side of history and hear stories of prisoners' crimes and experiences. … You'll hear about true and horrid crimes, including stories of bushrangers, murderers and pirates, and the brutal punishments of the day' (National Parks, 2021a, paras. 1, 2). While visitors are assured that this is not a ghost tour, they are told that 'tour content includes sordid tales and gore' (National Parks, 2021a, para. 3). For one travel site, this tour is for tourists who want 'to up the spook factor a notch' (Officer Travels, 2017, para. 15).

The second guided tour, 'Trial Bay Gaol: Life behind bars kids tour', invites visitors to 'Explore the gaol and discover its secrets on this action–packed family tour. … There'll be fun activities, games and treasure hunting. We'll use photo cards to locate special items and places. Come and play some old-fashioned German games, including Kegel bowling. Search for the missing gaol keys to earn your release!' (National Parks, 2021b, paras. 1, 3). The tour is recommended for ages 4–12, although any child under 16 must be accompanied by an adult. The last guided tour, 'Trial Bay Gaol twilight tour', focuses on providing information against 'the stunning natural backdrop at this golden hour' (National Parks, 2021c, para. 4). Visitors are told about the prisoners' experience of solitary confinement and how they were required to build the breakwall, how the German internees were treated and also stories 'about the villains and heroes who lived here, and the fate that met gaol escapees' (National Parks, 2021c, para. 3). The local wildlife is also heavily featured in marketing material as a drawcard for tourists. The site is also available for events and function hire, as well as upper primary school educational tours (National Parks, 2021d).

The coastal location, coupled with being surrounded by a national park landscape, makes this dark tourism destination similar in many ways to Port Arthur – a mixture of rural idyll and elements of dystopia. The 'ruins' of the prison 'hint at crenelated battlements, physical impenetrability, and violence through containment' while being set against a picturesque backdrop, bathed in sunshine for most of the year and offering a range of native fauna that delight tourists (Barnes and McIntyre, 2017, 59). The picturesque nature of the 'dark' site can create an unsettling effect for some visitors:

> I couldn't believe they would build a jail in such a beautiful location … a small mob of kangaroos making it their home. Wandering freely in and out of the gates, these roos can often be seen grazing next to the cells and sunning themselves by the old punishment room. It's eerie and cute all at the same time. (Officer Travels, 2017, para. 4)

In this site, some tourists have difficulty seeing the *darkness* of the site. Indeed, some tourists posting online emphasised that it was the only jail of the time that had no fatalities from illness or disease (Officer Travels, 2017), reinforcing the notion that the Trial Bay Gaol was not such a bad place to be. The tours on offer all seem to reinforce the 'lightness' of the site, offering tourists an entertaining, informative historical narrative. The camping grounds offer tourists a few distinct advantages: quick and easy access to the site, easy access to local surrounding communities and a dystopian ruined prison set against the 'stunning' Australian coastal landscape with native animals to enjoy.

## Courthouses

Courthouses offer the 'lightest' carceral tourism opportunities for visitors, simply because, unlike prisons, or even police stations, there are very few instances of violence or death. For Stone (2006), disused courthouses are dark dungeon sites – presenting places where 'justice' was enacted, and sentences of guilt or innocence passed. The previous chapter highlighted how the Beechworth Courthouse used light-hearted, immersive trial re-enactments, as well as soundscapes, to provide adequate atmosphere and narratives of real-life trials. School groups will often attend disused courthouses (in both rural and urban locations) on school excursions to re-enact past trials (or new ones).

Prior to exploring a case study of a dark tourism courthouse site in regional NSW, it is important to understand the historical context of the development of courthouses outside of urban centres. Before 1823, the court system in NSW constituted magistrates who were appointed by the colony's leading officers and settlers. The magistrates were required to deal with minor crimes, while indictable crimes were heard in the Court of Criminal Jurisdiction in Sydney Cove, with the Supreme Court of NSW established in 1823 (although it took until 1827 to receive a dedicated building in Sydney). In 1823, the New South Wales Act (Imperial) established the first Quarter and General Sessions courts and one district court in Sydney in 1823. Eight districts were provided with General and Quarter Sessions: County of Argyle, Bathurst, Camden, Cumberland (including Sydney), Gloucester, Londonderry, Westmoreland and St Vincent. For geographic regions more than 50 miles from these sites, a magistrate could be nominated from the area and, along with seven military jurymen, could hear cases of criminal misdemeanours. In the 1830s, there was a widespread shift in the way criminal justice was administered throughout the NSW colony. Settled regions within the colony were divided into police districts and police magistrates were posted in most regional areas dealing with minor violations of the penal code. The creation of 'local courthouses, gaols, post offices and police barracks formed the nucleus of small townships and laid the foundations of local government' (Wise and Roberts, 2016, 39). As such, in many rural and regional areas, courthouses are prominently placed and designed to invoke awe and respect.

## Berrima Courthouse

The Berrima Courthouse was designed during the 1830s, and finished construction in 1838, at a time when courthouses were seen as symbols of the importance of law and justice (Berrima Courthouse Trust, 2021a). As such, the Berrima Courthouse was designed to be formal and imposing. The Berrima Gaol (now the still active Berrima Correctional Centre) was opened in 1839 directly across from the courthouse. The courthouse was one of the earliest built in Australia (Western Herald, 1965), and the courthouse and gaol became the administrative centre of justice for the southern districts of NSW (Berrima Village, 2013).

The courthouse 'was used to convict criminals through to 1900 and saw crimes from cattle steeling [sic] through to notorious bushrangers and murderers' (Berrima Courthouse Trust, 2021b, para. 1). The first person to be sentenced to death at Berrima Courthouse was bushranger Paddy

Curran in October 1841 (sentenced and executed in the same month). The building ceased being a courthouse in 1884, and it was subsequently repurposed several times by the community.

During the 1920s, tourism from Sydney increased to the area in general, and 'operators at the disused gaol and courthouse enticed visitors with gory tales of convicts in chains and prisoner hangings' (Berrima District Historical and Family History Society, 2012a, para. 10). In 1965, the NSW state government invested money to renovate and preserve the 'authentic historical character of four early NSW courthouses ... [including] Berrima, Carcoar, Grafton and Bourke' (Western Herald, 1965, 2), all rural and regional locations. While the Grafton and Bourke courthouses were still active, both Carcoar and Berrima were being restored 'mainly for historical and architectural reasons, but will be extensively used for tourism' (Western Herald, 1965, 2). However, it was not until 1979 that a local group of trustees opened the courthouse as a museum to the public (Berrima District Historical and Family History Society, 2012b).

The Berrima Courthouse is owned by the NSW State Government, yet it does not receive any local, state or federal funding. Instead, it is a registered charity and 'it's [sic] conservation programs are totally funded by visitors and profit from retail sales in the "Petty Jurors" book and souvenir shop' (Berrima Courthouse Trust, 2021c, para. 1). As such, tourism is ensuring the continuity of this site:

> Your entry fee to tour the courthouse is a real and measurable way of assuring that this historic icon is preserved and cared for now and into the future. We thank you for helping us to preserve this grand building. (Colin Gelling (CEO) cited in Berrima Courthouse Trust, 2021c, para. 3)

Owing to the reliance on tourism, the narrative presented (and importantly *how* it is presented) becomes crucial. While receiving no funding, it is still owned by the government and, as such, will be guided by political ideologies and narratives. At the same time, the lack of official funding means that sites like the Berrima Courthouse need to structure their activities and storytelling narratives in a way that appeals to a diverse range of people to maximise tourism potential (and to entice people from urban areas to travel to the site). As such, they need to cater to those who want to know more, but in a light-hearted, family-friendly environment, as well as those tourists that want gory details and titillation.

There are three types of tours available at the courthouse: self-guided tours, offered 364 days a year, guided tours for groups and schools, and ghost tours led by an external company which are advertised by the courthouse as 'Become the ghost hunter for the night' (Berrima Courthouse Trust,

2021a) to experience 'ghosts, murders, hangings and twisted tales' (Berrima Courthouse Trust, 2021b, heading). The self-guided tours allow tourists to walk through the courthouse at their own pace. However, each tour has a set 'start-time', which is set at every 30 minutes to allow visitors to experience the 'Captain Starlight Theatrette' (Berrima Courthouse Trust, 2021d). There is another presentation, 'Treachery Treason and Murder', which is an 'entertaining sound and light show' held in the grand courtroom (Berrima Courthouse Trust, 2021b, heading). The courtroom hosts an 'immersive ... experience' utilising mannequins and a light and audio show (Berrima Courthouse Trust, 2021b, heading) to re-enact 'one of the most notorious cases put on trial' (Berrima Courthouse Trust, 2021e, para. 2).

The Australian Paranormal Phenomenon Investigators (APPI) Ghost Hunts and Tours run several night-time ghost tours and claim that the Berrima Courthouse is 'one of the most haunted buildings in Australia' (APPI, nd, para. 6). Tourists can choose from an 'immersive' tour, an all-night tour (running from 8 pm to 6 am), Halloween tours (running from 8 pm to 9.30 pm) or a parapsychology ghost hunt (running from 8 pm to midnight). According to the AAP Ghost Hunts & Tours website (nd), during the interactive 'Immersive ghost tour', tourists will learn the history of the site, see footage of past haunting occurrences, as well as participate in vigils to try to connect to spirits. Children aged over 12 years can participate in these tours and also in the Halloween ghost tour (last offered in 2020), which encouraged tourists to dress up in their best Halloween costume and promised a 'few SHOCK scares to fit the theme' (Sticky Tickets, 2020, para. 7).

The parapsychology ghost hunt advertises that the Berrima Courthouse was the 'site where convicted murderers were told they would hang' and was 'just across the road from their final place of death' providing 'tragic history and several different ghosts who go with it' (APPI, nd, para. 9). The tour invites tourists to participate in a range of experiments to study the psychic phenomena present at the site.

The all-night investigation promises tourists the ability to walk through the 'historic precinct', hear stories and hopefully 'make contact with any of the spirits who may still remain' using technical equipment and 'some spiritual methods' (APPI, nd, paras. 13, 14). While this is an all-night tour, no sleeping quarters are advertised, and there is no information about bringing along sleeping equipment, suggesting that the investigations carry on all night.

Tripadvisor ranks the courthouse as 4 out of 5 (from 122 reviews), and the third 'top' tourist activity for Berrima. Visitors seem to find the museum educational and fun but also 'corny', 'cheesy' and 'spooky' (Tripadvisor, 2021c). One visitor (featured on the website) wrote 'Fantastic museum! ... fascinating facts and extreme silliness ... finally the "spooky light show" at the end is a hoot, hardly terrifying but great fun and a good way to visualise a

Georgian court' (Berrima Courthouse Trust, 2021b, first quote of reviews). As mentioned, the museum also offers tourists the opportunity to purchase a piece of their trip from the gift shop the 'Petty Jury Bookshop', which 'specialises in local and Australian history; local townships, first fleet posters, convict history, souvenirs and postcards' (Museums and Galleries of NSW, 2021, para. 5). In the past, the museum offered the opportunity for children to 'dress up'.

In essence, the Berrima Courthouse is a dark fun factory for most of the tours– the emphasis is on fun, entertainment and titillation. Even at night, the focus on parapsychology is not necessarily a form of respect – it is to 'make contact with any spirits that remain'. As such, while the site is informative and educational, the emphasis is on creating a fun-centric environment. Returning to the fact that the site receives no government funding, it may be that this is the model that *needs* to be adopted to attract continual visitors. The notion of hosting themed events, such as the Halloween night, is certainly adopted by numerous tourist sites in urban areas, so perhaps such themed events attract both local and 'out-of-towner' tourists.

## Police museums

Police museums 'can be conceptualised as "dark tourism" locales that depict death and suffering' (Ferguson et al, 2019, 318) while informing the public of police history by utilising exhibitions of artefacts relating to policing. While police museums are widespread across the world, they differ substantially in size, resources and popularity compared to penal museums, which are usually larger, better funded and well-known.

Police museums are also often located in decommissioned police stations, watch houses or within police headquarters. According to Ferguson et al (2019, 318) police museums 'are one of the ways police communicate about their past and present practices in an attempt to foster public reassurance and legitimacy'. While most police museums focus on education, some adopt a more interactive element to entertain tourists. For example, the Queensland Police Museum, located on the ground floor of the Police Headquarters in Brisbane, Australia, offers tourists 'a confronting mock up murder scene to solve' (Panayotov, 2023, para. 2).

Like penal museums, police museums are often accused of presenting biased exhibits that memorialise police work and officers killed in the line of duty, as was evidenced in Chapter 4 with the Victorian Police Museum. In many museums, the role of the police in inflicting pain and detention is overlooked or downplayed. Instead, the focus is on 'suspect populations' that 'deserve' or 'require' detention or violence as a part of reinforcing state control (Ferguson et al, 2019, 320). While research has investigated this internationally, there is little information available on police museums in

Australia, and in particular how rural or regional policing is portrayed (or overlooked) in such spaces.

Within Australia, each state and territory has at least one police museum. In most parts of Australia, these sites are based in urban areas, such as Sydney (NSW, the Police and Justice Museum), Brisbane (Queensland, Queensland Police Museum), Melbourne (Victoria, the Victoria Police Museum), Canberra (Australian Capital Territory, the Australian Federal Police Museum, which is currently closed to the public), Hobart (Tasmania, the Tasmania Police Museum), Adelaide (South Australia, the South Australian Police Museum) and Darwin (Northern Territory, the Northern Territory Police Museum, also currently closed to the public). Western Australia is the only state not to have a police museum in its capital, however, the state, as well as the Northern Territory, have a number of rural and regional police museums. For example, in the Kimberley region of the Northern Territory, there is the Timber Creek Police Station and Museum, as well as the Borroloola Police Station Museum. Western Australia has three rural and regional police museums, including The Old Police Station Museum in Mount Barker, the Bridgetown Police Station Museum and the Morawa Museum and Old Police Station.

Visiting police museums is often more difficult than in other carceral sites (regardless of their location), with limited opening hours, and some only accessible by police officers. For example, the Tasmania Police Museum is only open on Tuesday mornings between 10 am and 1 pm, and the Police and Justice Museum in Sydney is only open on weekends and every day during school holidays. The opening hours sometimes relate to necessity – for example, in Tasmania the museum is run by volunteers, limiting the available times to have the site open. For those museums located in rural areas, visitors must travel considerable distances from the nearest major city. For example, Timber Creek Police Station and Museum is a six-hour drive from Darwin. Similarly, the Borroloola Police Station Museum is over 11 hours by car from Darwin. In Western Australia, driving from Perth, visitors would need to travel over four hours to Mount Barker, two hours to Bridgetown and almost four hours to Morawa. For these museums, having reliable opening times is crucial, as visitors who travel to find it closed leave poor reviews on sites such as Tripadvisor. For example, one traveller from the Australian Capital Territory attempted to visit the Timber Creek Police Station and Museum based on its reputation only to find it closed with no information about how to access it (Tripadvisor, 2023).

Those museums located in rural and regional areas tend to focus on the history of the region – providing more localised exhibits and narratives, thus showcasing more of the idiosyncrasies of rural policing. Some have local tour guides, while others allow tourists to visit unaccompanied.

While urban police museums tend to focus on the city they are housed in, efforts have been made to highlight specific 'rural' aspects. For example, the Tasmania Police Museum is currently embarking on 'a project to identify, record and photograph as many old police stations in Tasmania possible', with particular emphasis on those located in rural Tasmania (nd, para. 1). Many urban museums also focus on bushranger and gold rush stories, as well as early interactions with Aboriginal people and the employment of native trackers, which, while not necessarily focusing on the rural element, inevitably highlight the rural to a limited extent. The Police and Justice Museum in Sydney also showcases mugshots taken from rural and regional gaols, and thus provide some information on the offenders and crimes occurring outside of urban centres.

Plans are also underway for the creation of the Museum of Australian Policing, which will be located in Canberra and is set to open in mid-2024. The museum, to be funded by confiscated proceeds of crime, will be the first museum of Australian policing in the country and will:

> showcase the great work police do from Albany in Western Australia to Coffs Harbour in New South Wales, from Dover in Tasmania to Maningrida in the Northern Territory. It will also feature some of the amazing work Australian police have performed across the globe. (AFP Superintendent Dean Elliott cited in Byrnes, 2022, para. 3)

As such, there appears to be a recognition of the need to showcase more rural and regional policing endeavours within this museum. How rurality is portrayed or analysed will be interesting to see.

Following other carceral tourism sites, many of the police museums across Australia offer tourists the opportunity to take home a memento of their visit from their gift shop. Many stock 'crime' related books, while others offer more 'fun' 'gifts', for example, visitors to the Police and Justice Museum can buy a 'mugshot height chart', which allows you to:

> Create a convict-style case file on the growth of your little terrors, complete with photographic evidence!
>
> The Suck UK Mugshot Height Chart is a double-sided paper poster and set of 10 number cards. Start processing your usual suspect by ensuring the poster correctly records the height, then ask the perp to hold the numbered age cards in front of their chest. Take two photos, one front-view, one side-view, and you'll have an almost official criminal record to add to the photo album. (Museums of History NSW, 2023, paras. 1, 2)

Other items available from a range of museums are stubby holders, tea towels, magnets and even 'crime scene' bracelets.

# Conclusion

Carceral tourism in rural and regional Australia has been highly influenced by a commercial ethic. While each site discussed in this chapter has clear historic and heritage value to the understanding of Australia's past carceral practices, each site has ultimately been preserved and managed because of the tourism interest in crime and justice. Families and schoolchildren are encouraged to visit each of these sites, influencing the tourism infrastructure and opportunities. As such, in most of the sites, the day tours are geared towards family-friendly entertainment, resulting in the site sitting towards the lighter side of Stone's spectrum, and being a mix of a dark dungeon and dark fun factory. Yet at night, many of these sites transform themselves into selling more 'dark' and macabre events, moving the site further along the darker scale. While the material itself is darker (hangman tales and ghost tours), the presentation of the material still ensures that the site remains a dark dungeon – the site aims to entertain and provide thrills at an authentic site.

One question then arises: to what extent do these rural and regional sites misrepresent the institution and its former residents? It is clear that the material being presented as 'authentic' is very much staged, yet this does not necessarily mean that it negatively affects visitors' (educational) experiences of the site. From reviews on Tripadvisor, it is clear that many tourists have left each of the sites with a deeper understanding of the history of the site and past criminal justice practices more broadly. Further, there is a sense of empathy for the prisoners (except at police museums). At the same time, some of the sites have relied upon selling tactics, such as relying on celebrity prisoners or gruesome tales to entertain and titillate tourists. Like previous studies, there was little coverage at any of the sites about more 'serious' material such as suicides (except for Ararat, where hearing stories of 'suicides' is actually promoted on their webpage) or guard and inmate brutality. Instead, tourists are offered the 'lighter' side of the criminal justice system, and in some of these places, the level of empathy that visitors feel for prior prisoners is indeed questionable.

As mentioned at the start of this chapter, carceral tourism provides tourists the opportunity to visit sites that have been responsible for inflicting physical and emotional pain on countless individuals (sometimes causing death). The entertainment focus of such sites, allowing visitors to be theatrically arrested, locked up and threatened with execution raises a host of moral quandaries surrounding the transformation of these spaces into spaces of infotainment, as does the general approach of most sites to disengage with discussions relating to minority groups or 'unsavoury' topics such as suicide or guard brutality. For example, Witcombe (2013, 158) questioned the ethics of dark tourism strategies that allow the visitor to 'roleplay' a position that 'they can never take on and which ultimately provide entertainment for themselves at the cost of a real understanding of what might have led real people to experience the horrors of the' site in question.

Most visitors scarcely question the ethics of visiting a decommissioned site: indeed, sites tend to provide tourists with a sense of moral superiority compared to the uncivilised past that they are witnessing. That is, carceral tourism sites often provide the visitor with a 'sense of moral progress' within the criminal justice system by presenting older, more 'barbaric' practices, such as public executions, as being morally reprehensible (Welch, 2013, 491). Further, the 'othering' of prior inmates and focusing on their crimes and violence allows the visitor to disengage in any moral quandary associated with being present at such a site or the questioning of continuing criminal justice practices such as incarceration.

Despite this, there are many positive aspects of penal and jail museums. As Strange and Kempa argue:

> preserved prisons are stony silent witnesses to the things former regimes were prepared to do to people who violated laws or who seemed threatening or suspicious. The murkiest project of all would be to close them to tourists rather than to confront the ongoing challenge of interpreting incarceration, punishment, and forced isolation. (Strange and Kempa, 2003, 402)

This is equally true of courthouses and police museums. Each of the sites recognised its important role in presenting the history of carceral institutions in Australia – especially as so many other buildings have not been preserved. In addition, many of these sites have served their communities in a variety of ways and continue to serve as an economic boon to the region.

All of the prison and courthouse sites had their own café and gift shop – albeit to varying levels of consumerism. The consumption of criminal justice-related tourism starts with the purchase of an admission ticket (or entry by donation), and:

> Once inside, trinkets for purchase at the end of tours also encourage a memorializing of the site. ... The locks for sale allow the tourist visitor to remove a quasi-relic from the memorialized site and use it as a cue in creating their own narratives concerning imprisonment and punishment once visitation has ended. Handcuffs, whistles, tin mugs and various forms of attire serve the same function. Pretes (2002) refers to this process of selling items related to the tourist site as 'mining the tourists'. (Walby and Piché, 2011, 461)

The variety of 'trinkets' at each site varied considerably – some focused on selling books and historical information, while others opted for selling local wares such as soaps. Others did provide the more 'trinket' type souvenirs such as convict fridge magnets and Ned Kelly souvenirs. The level of corporate or state-level government involvement in the site probably has a significant role in the level of tourism infrastructure attached to the gift shop. From the sites covered in this chapter, it appeared as though those sites owned or managed by volunteers or national parks management offered fewer 'trinkets', while more 'organised' or 'corporate' sites offered a wider range of consumable souvenirs.

There are numerous differences between urban and regional dark tourism carceral sites. One key difference relates to ease of access – urban sites are more likely to attract tourists simply because they are 'there'. In contrast, in rural and remote areas, communities need to advertise broadly and also ensure that the whole region has enough tourism opportunities available to attract tourists. A second difference is the landscape and atmosphere of a site. In urban areas that are surrounded by city sounds it is easy to remain in the 'present' and not truly understand what life 'inside' may have entailed. Sites in rural and regional areas without this excessive or constant noise, and with more space to really 'showcase' the site, offer a unique advantage once tourists are physically there. This can also distort the history of the site, as seen at Trial Bay Gaol, where the sheer beauty of the landscape challenges tourists' feelings that the site was a gaol. Here, the rural idyll remains dominant in this landscape, to the point of diminishing previous prisoners' experiences of being incarcerated, arguing it would have been a holiday, easily forgetting the loss of freedom and imposed control over these people. A third difference relates to the level of 'lightness' presented at the sites and the selling of macabre in a thrill-seeking manner.

One hypothesis is that some of these rural/regional sites *need* to rely on dark fun factory tactics such as ghost tours to ensure a wide tourist base. For example, as noted in the discussion of Ararat's J Ward, they attract international and national ghost hunters and paranormal investigators, whereas normally most tourism to similar areas only attracts a domestic audience. The case of Ararat also highlights that sometimes a corporate model may not be in the best interests of the community or the site itself. The continuation of a volunteer-run business receives positive reviews and provides the community with ongoing ownership over the site and how it chooses to 'sell' itself. The ownership of the site by ACRE now also serves to support new community endeavours, ensuring the site remains a significant building within the community.

The original rurality of the location itself also dictated that these towns received a criminal justice organisation. The Berrima Courthouse was built to administer justice more swiftly to a region geographically distanced from Sydney. Similarly, all of the prisons were built in regional areas because transporting convicted offenders back to the city was prohibitive. Trial Bay Gaol was built specifically for its rurality and the need for labour to build a breakwater. The inherent 'rurality' of these locations shaped their original construction and philosophies and continues to have an impact on how they 'sell' their product to tourists today.

6

# Serial Killers and Sensational Crimes

Travel to sites associated with serial killers (whether that be burial sites, sites of the murders or even simply where the serial killer lived or visited frequently) can be another popular tourist activity. Considering the widespread fascination with serial killers within popular culture (television shows, movies, true-crime books and so on) the popularity of dark tourism sites focusing on serial killers is unsurprising. Demonstrating societies' widespread awareness of serial killers, Kincaid stated:

> Recent surveys of the store of general knowledge possessed by Americans reveal that 11 percent have a firm grasp of evaporation; 23 percent know pretty much where the equator is. ... Yet a solid 100 percent, every single adult and child, knows Jeffrey Dahmer, identifies him as a serial killer, homosexual, cannibal, ghoul. (Kincaid, 1997, ix)

Kincaid (1997, ix) asks why this is the case and, in particular, 'what cultural itches do they [serial killers] scratch?'. Whatever the attraction, tourism associated with serial killers has a long and rich history.

By way of example, the murder sites of Jack the Ripper, whom Gibson (2006, 52) has referred to as the 'serial killer superstar of all time', immediately attracted sightseers and continues to be a popular tourist attraction within modern-day London. At the time of the murders, police were required to force their way through the 'morbid curiosity seekers' in order the reach the body of Annie Chapman (Wilson and Odell, 1987, 22). Some of Chapman's former neighbours:

> were able to cash in on Annie's misfortune by charging sensation seekers a few pence each for a look into the yard where she was done to death. (Wilson and Odell, 1987, 24)

Those lucky enough to have a view from one of the many buildings surrounding the sites sold window seats, and there was no lack of customers. The streets leading to the murder sites were literally choked with thousands of people. (Gordon, 2001, 116)

The murder scene of Annie Chapman not only saw spectators but also vendors setting up stalls to sell fruit and refreshments to the spectators (Begg, 2005). Here we see elements of what we know of today as 'dark tourism' with 'tourists' paying to see a site of death and having the potential to have a 'guide' (acquaintance of the murder victim) providing explanations for spectators as early as 1888.

Another example of impromptu dark tourism that became commodified and 'sold' relates to Jeffrey Dahmer, the notorious American serial killer who murdered 17 victims between 1978 and 1991. Soon after the stories made international headlines, the Oxford Apartments in Milwaukee, United States, where Dahmer killed and stored many of his victims 'became a tourist trap, attracting curious visitors from as far away as Japan' (Fox and Levin cited in Gibson, 2006, 565). The building was demolished in 1993, 15 months after his arrest, in an effort to deter tourism. However, the demand for tourism activity led to the Shaker's Cigar Bar establishing a controversial Cream City Cannibal walking tour in 2012 (now offered through Hangman Tours), despite protests from the victims' families and the local community. The tour's sensitivity and morality have been questioned:

But questions of morality begin even before the tours do. Before setting off, participants are encouraged to buy a $20 Dahmer T-shirt emblazoned with his jail mugshot on the front and 17, the number of young men and boys he killed, and his name on the back, sports jersey style. As the tour proceeds, weaving between Indulgence Chocolatiers and La Cage, guides provide alarmingly specific details bout [sic] Dahmer's murders: the weapons used, his necrophilia and dismemberment of the victims' bodies, how he disposed of the remains. (Miller, 2019, para. 3)

The tour was featured in the Netflix series *Dark Tourist*, heightening awareness of the tour, and adding a 'selling' feature for the company. It has been suggested that most of the participants on this tour are 'out-of-towners', suggesting that people have travelled to the site (Miller, 2019). Whether this travel is specifically for the site, or just an 'add-on' while in the city is unknown. Regardless, the success of tours such as these, despite local opposition, highlights the ongoing demand for serial murder-related tourism.

At the beginning of the book, the motivations for tourists travelling to 'dark sites' were briefly discussed. For Lennon (2010, 217), when discussing sites of criminal activity 'the appeal of such sites is less to do with the perpetration of an illegal or criminal act but rather a dark fascination we appear to possess with evil and the acts of evil that we can perpetrate'. That is, tourists are not interested in visiting all types of crime (for example, sites of theft or fraud), rather, the main attraction is the level of tragedy and suffering that has transpired.

While there are varying levels of tourist activities at sites of torture, death and tragedy associated with sensational crimes, some sites can be characterised as 'dark shrines' with a greater focus on remembering and reflecting on the atrocities, while others with a more touristic element tend to focus more on 'entertainment', with stories of gore and merchandise for sale. For Stone (2006, 155), dark shrines 'essentially "trade" on the act of remembrance and respect for the recently deceased'.

Internationally, there have been several rural dark tourism sites associated with serial killers, although they seem to be more likely to be based in urban areas. As discussed in the introduction, the farmhouse of Edward Gein in Plainfield, United States, became a tourist destination, as did the residence of Fred and Rosemary West in Gloucester, England. Rural tourism sites of serial murders and other sensational crimes have additional layers of complexity to consider. Travel to the sites is often incorporated within a larger trip or, in some cases, it may be the specific purpose of the travel. Yet planning tourism to these areas often requires forward planning and an intention, rather than simply seeing it advertised on the day and deciding to go. Further, the level of infrastructure around these tourist sites can range considerably – from the dark shrines outlined by Stone (2006) to more formalised tours or, in the least, places of businesses selling souvenirs. Three of Australia's most well-known sites of serial and mass murder are considered in this chapter to tease out some of these rural considerations: Belanglo State Forest (Ivan Milat), Snowtown (Bunting, Wagner, Haydon and Vlassakis) and the Broad Arrow Café at Port Arthur (the site of the mass shooting committed by Martin Bryant).

Historically and in modern times, the Australian outback has come to be portrayed as a space that is dangerous for urban people and, in particular, tourists (Rosser, 2013). Following on from bushrangers, rural Australia continues to be negatively associated with death, tragedy and torture with some of Australia's most sensational and horrific murders occurring in remote areas of the country. The rural landscape plays a large role in the allure of the (serial or mass) killer where the landscape itself becomes sinister and threatening. Murderers become labelled by the media as monsters, and the rural idyll is frequently inverted into a rural dystopia when serial and/or mass murderers are portrayed. Further, the rural 'allows' serial killers to become monsters:

The serial killer is frequently portrayed as the 'everyman', blending in with society until their monstrousness is revealed to their victims.

In a space such as the outback, however, there is no need to conform to an acceptable social image, and they are permitted to behave as monsters. In this way the geography of the outback, and the national memory of the events that have occurred there, creates a place for monsters. (Rosser, 2013, 178)

As such, rural idyll notions of beautiful landscapes populated by communities with wholesome values, kinship ties, 'religious piety and social stability' are swiftly replaced with perceptions of death, macabre and monsters, and this 'stain' remains for much longer than any urban area would be tarnished for because 'rural dystopias are marginal spaces, beyond the civilising qualities of the urban' (Rofe, 2013, 269–270). These rural locations continue to stand out in our national identity – forever associated with death and horror.

## Belanglo State Forest, New South Wales

Between the period of December 1989 and April 1992, seven backpackers disappeared on four separate occasions in Australia – in three cases there were two backpackers travelling together. One of the common elements between the hitchhikers was that they were known to hitchhike in order to travel to different parts of Australia. Another commonality between the disappearances was that they were all hitchhiking between Sydney, NSW and Canberra, Australian Capital Territory along the Hume Highway, NSW Australia. The landscape along the Hume Highway between the two cities is mostly bushland, with towns spread between half an hour and an hour apart (with some towns bypassed by the highway). As such, the route transverses many landscapes that are isolated and uninhabited. The victims were Deborah Everist and James Gibson from Victoria, Simone Schmidl, Anja Habschied and Gabor Neugebauer from Germany, Joanne Walters from Wales and Caroline Clarke from England. Their bodies were discovered between September 1992 and November 1993 in the Belanglo State Forest, south of Berrima in the Southern Highlands of NSW.

Following the discovery of the bodies in Belanglo, 'the backpacker killings became the subject of intense international media scrutiny and speculation' with headlines such as 'Beast of the bush: Brit girls victims of Oz serial killer' (McGowan, 2019, para. 3). Over 30 years later, the 'murders still haunt Australia … and Belanglo has become a byword for horror', with questions still unanswered about possible further victims and whether Ivan Milat had an accomplice (McGowan, 2019, heading and para. 6).

Consequently, Belanglo Forest is now primarily known as the killing and burial site for serial killer Ivan Milat (Twyford-Moore, 2017). Although Ivan Milat lived in Sydney at the time of the murders and picked his victims up in Liverpool (an outer suburb of Sydney), his crimes are synonymous with rural NSW. The rural outback has become the site of inherent danger, the rural dystopia. As such, it is within the isolated forest that a memorial to the victims has been constructed and at least one tour company has attempted to run tours.

A memorial was dedicated to the seven victims of Milat on 5 February 1994 (see Figure 6.1). The original memorial consisted of a plaque attached to a rock at the juncture of the Northern Firebreak Road and Belanglo Firebreak Road, along the Western Boundary Walk in Belanglo State Forest. Simone Schmidl's family have also erected a cross bearing her name next to the official memorial. The memorial can be likened to Stone's notion of a 'dark shrine'. Photos of the memorial show a myriad of objects left at the site: necklaces hung over the stone bearing the plaque, statues (one an owl, another a guardian angel), a beer bottle propped up against the stone, seashells, painted rocks/stones and flowers. Simone's cross is adorned with a scarf wrapped around it, with floral tributes placed at the base (Atlas Obscura, 2021). The main motivation to visit this memorial would be to remember the victims or from curiosity – there is no tourism infrastructure offering tourists refreshments, shelter or information.

This vacuum of tourism potential was filled in 2015 when Goulburn Ghost Tours attracted considerable negative media publicity over the introduction of Ivan Milat themed ghost tours within Belanglo Forest. The tours began in late June 2015, and by mid-July the tours had been cancelled. The tour cost $150 and the Goulburn Ghost Tours website promoted the experience as an 'extreme terror tour'. Further, the website invited tourists to:

Come with us to Belanglo where Ivan Milat buried the bodies of his victims!
Once you enter Belanglo State Forest you may never come out ...
Are you ready to turn grey overnight?
Do you love to be frightened?
Would you like to solve a crime? ...
Are we being followed ... watched?
Is there another victim waiting to be found? (Goulburn Ghost Tours, 2015, paras. 1–3, 7)

At the end of the webpage, in very small font, there was an 'important notice' section outlining 'all tours are conducted with the greatest respect given to victims and their families. Please remember these crimes were committed over 20 years ago' (Goulburn Ghost Tours, 2015, para. 16). This message contradicted the overall tone of the rest of the webpage, which promised tourists a thrilling and detailed evening.

A few weeks after the tours had been active, there was considerable public and political outrage. For the then chief executive officer of the NSW

**Figure 6.1:** Memorial to Ivan Milat's seven victims, NSW, Australia

Note: This image highlights the 'rurality' and isolation of the site.

Source: Wise, N. (nd), 'Memorial in Belanglo State Forest'

Victims of Crime Assistance League, Robyn Cotterell-Jones, the tours were insensitive to the families of the victims:

> 'It will be greeted with revulsion and disgust with people who would like a bit more respect for their own suffering,' she said.
>
> 'While human beings seem fascinated by the macabre and frightening, for the families of victims, the impact of the death of their loved ones is never ended. For them, to hear people are using places of such horror for their amusement and profit is obviously going to cause scars to be ripped open again.' (cited in Patty, 2015, paras. 12, 13)

One concerned community member started a 'stop the Belanglo State Forest Ivan Milat Terror Tour and make Goulburn Ghost Tours issue an apology to the families of the victims' petition on the change.org website (change.org, 2015, heading). The petition was signed by 12 people before the tours were closed.

The overt media attention led to Goulburn Ghost Tours spokesperson, Louise Edwards, appearing on the popular program *The Project* and providing statements for various media outlets. She told *ABC News* (2015) that the tour was sensitive to the victims while filling a niche gap for those interested in paranormal activity at crime scenes:

'It comes down to a moral predicament, it's an argument that nobody can win because it comes down to opinion.'

'We go to the memorial site and we pay our respects to Ivan Milat's victims,' she said.

'There is absolutely no-one jumping out of the bushes, we don't have recordings of people screaming. It's absolutely not that kind of tour.'

'We don't say any of the victims' names and say "are you here?"'.

'We run the tours with great respect.' (Louise Edwards cited in ABC News, 2015, paras. 39, 41–44)

Edwards also cited the time between the killings and the tour, stating that with 20 years passed, and so much information being publicly available about the crimes, the company did not anticipate the level of controversy or for the victims to be impacted negatively (ABC News, 2015). Edwards, and the tour company, received criticism on *The Project* (a commercial television news program) and via viewer comments on social media accounts. Facing this criticism, it was announced live on *The Project* that the tours would not be held anymore, despite previous tours selling out, and forthcoming tours already being booked.

The day after *The Project* interview aired, it was reported that the company had removed the contents of their webpage and replaced it with the following message: 'Hang tight, we are now looking for a tour with out [sic] such an extreme response' (Thompson, 2015, para. 8). At the time, there was speculation as to whether the company would survive the controversy, with market and public relations experts consulted by the media to ask about the viability of such a small company after a breach of public confidence (Thompson, 2015). Interestingly, the question of location was not raised; that is, the possibility that the company was based in regional NSW while being featured in the media, was not considered in the analysis of whether the company could, as Thompson (2015) framed it, 'recover from the incident'.

Within the media coverage, particularly through *The Project*, there were comparisons made between the Belanglo Tours and international serial killer tours such as Jack the Ripper and Jeffrey Dahmer, with *The Project* asking, 'so where is the line between historical tourism and exploitation?' Similarly, public relations expert Janey Paton believed that 'They took it a bit too far given such tragic events did occur, there's a fine line between a historical exciting tour and scaring the hell out of your customers' (cited in Thompson, 2015, para. 12). One possible explanation for the 'suitability' of a tour is the potential to generate income for a community. Belanglo is a state forest isolated from communities and is over a one-hour trip from Goulburn where the tours originated. Similarly, Goulburn is not usually associated with the murders (despite Ivan Milat being housed in the maximum-security prison in Goulburn for much of his incarceration), and as such, it may seem more reprehensible for the company to be making money from atrocity without a clear economic benefit to the community.

## Snowtown

Rofe (2013, 270) argues that the case of Snowtown is the 'embodiment of the rural dystopia and a site of dark tourism'. Snowtown is located approximately 150 kilometres north of Adelaide in South Australia and is a small rural town known for farming and agriculture with a population of 467 in 2016. In 1999, the town became well known locally and internationally as the site for the 'bodies in barrels', and what is now known as Australia's worst serial killings.

On 20 May 1999, as part of a year-long police operation named 'Chart' into three missing persons, more than 50 police from Adelaide searched the vacant State Bank branch building in Snowtown. Residents of Snowtown told police that the bank vault was rented to John Bunting to store kangaroo carcasses (Koehler et al, 2009). At the time, the media reported that the police were not expecting to find human remains (Debelle, 2019). However, upon entering the bank vault, the police found the remains of eight bodies stored in six 44-gallon plastic barrels filled with acid. John Bunting, Robert Joe Wagner, Mark Ray Haydon and James Vlassikis were convicted of the murder of 11 victims spanning 1992 to 1998 (although there were 12 deaths associated with the group). The victims were Elizabeth Haydon, Clinton Trezise, Ray Davies, Suzanne Allen (the jury was unable to convict in this case), Michael Gardiner, Thomas Trevilyan, Gavin Porter, Troy Youde, Frederick Brooks, Gary O'Dwyer, David Johnson and Barry Lane. All of the victims were known to the assailants.

The media were quick to report on the case, and Snowtown 'achieved international notoriety almost overnight' (Rofe, 2013, 265). Within six days, the serial murders were known as the 'Snowtown Murders', and the perception of the town quickly changed from the rural idyll to a rural dystopia:

> There was a time, no doubt, when the name of the sleepy South Australian hamlet evoked a delicious shiver, conjuring up images of snowmen, clean crisp air and warm welcoming fires.
>
> It still evokes a shiver today, but of an entirely different kind; a spine-chilling shudder of revulsion. And the images it throws up are not of snowmen but of monsters, of foul air heavy with the stench of dismembered and decomposing bodies. (Smith, 2002, 34)

As Rofe (2013, 266) outlines, journalistic embellishments such as these led to the 'explicit vilification of an entire place and its community', essentially turning the geographical location into a dystopia.

The subsequent media coverage of Snowtown played a strong role in the growth of dark tourism by generating awareness, interest and appeal. Even though none of the four convicted perpetrators or victims were from Snowtown, and only one of the murders actually took place in the town itself, Snowtown became a popular dark tourism site 'where residents cashed in with guided tours and hasty garage sales staged for the morbidly curious' (Newton, 2006, 243). A newspaper piece reported that 'ghoulish tourists' had Snowtown 'in their sites' as visitors began to arrive the weekend after the police had made the discovery (Cock, 1999, 6). Goodfellow (2003, 12) elaborated on the tourist element by stating that bus tours and interstate tourists came to take photographs and 'sniff' the stench at the bank. A 'dead bodies' sticker also appeared on the Snowtown sign on the town boundary under the heading 'Welcome to Snowtown' (Weir, 2000, 30). According to Dalton (2015, 183), Snowtown is an example of 'undesirable crime-related dark tourism', and indeed one local referred to tourists as 'terrorists' (Goodfellow, 2003, 12).

According to Phillips (1999, 12), 'the torrent of thrill-seeking tourists' had 'reduced to a trickle' by August 1999. Yet, the community was very much affronted by the behaviour of tourists during this time:

Why, asked a disgusted secondhand goods dealer, Mrs Thelma Drew, would a tourist embarrass herself by bending to her knees to smell the bank's external air vents? And why, she demanded incredulously, would two elderly men from Port Arthur, of all places, make a pilgrimage to Snowtown to 'have a look'? ... the tourists who had flocked into town asking 'stupid, ridiculous questions' would not let anyone forget. 'We are getting on fine,' Mrs McCann said. 'But gory people keep asking us what was in the vault. Did we see it?' (Phillips, 1999, 12)

According to one resident of Snowtown, Barry Drew (cited in Coleman, 2000), tourists were stealing pot plants from the bank's front garden and fruit off the trees at the back of the bank as a way of taking a memento from the dark site. As a result of this behaviour, Drew began to sell fridge magnets, caps and T-shirts that displayed slogans like 'Snowtown? you'll have a barrel of fun' and 'I've been to Snowtown and survived' in his second-hand store (Coleman, 2000). In a relatively short period, the shop had sold over 100 magnets (at a cost of $2.50 each), and they had received national and international inquiries about merchandise relating to the murders (Wakelin, 1999, 6). Drew, who was one of three shop owners to sell merchandise, believed that the selling of the magnets stopped the theft of items around the bank.

However, the selling of 'macabre' and 'kitsch' items also led to community conflict, with many of the older residents of Snowtown wanting tourists to come to the town to enjoy art and community festivals (Coleman, 2000). In addition, the mother of victim Barry Lane, Mrs Sylvia Lane, told the press that she found the magnets 'distressing' and 'sick' (Wakelin, 1999, 6). In response to the negative media, Mr Drew was quoted as stating 'If it had been in the US they would have been running barrel races by now. ... We've held off for over three months', and another seller stated that they sold the humorous magnets as a way of making a joke, which is 'part of the healing process' (Wakelin and Oakley, 1999, 4). Edward Gein's crimes similarly attracted 'gallows humour' with jokes, riddles, puns and rhymes being created. A psychologist studied this and argued that 'such humour served as a psychological defense, a way for people to confront their anxieties without actually admitting their fears' (Foote, 2003, 209).

With increased media attention and continuing tourism, the community explored different options for 'dealing' with the site of the bodies. In 2000, the town held an inaugural Snowball and Snow Festival to 'help shake the image left by the bodies-in-the-barrels killings' (Advertiser, 14 August 2000, 7). There were suggestions to demolish the site, even at the local government level. However, after asbestos was discovered, the demolition costs were prohibitive (Rofe, 2013).

The bank itself has been sold several times since the discovery of the bodies. In November 1999, a few months after the bodies were found, the owners of the bank (who were local farmers in Snowtown) decided to sell the building and the attached four-bedroom house. The owners told the media that they hoped that whoever bought the property would convert it into a ' "tasteful" tourist attraction' because the bank would remain a landmark of the historical event and it could be used to benefit the town by increasing tourism (Huppatz, 1999, 13). The building was bought at the end of 1999, and the new owner intended to turn the site into a healing centre.

By 2003, the site remained a private residence and the town itself was in economic decline. Journalist Maguire (2003, 16) wrote that Snowtown 'looks as sorry a place as you would find, with shopfront after shopfront displaying little more than cobwebs and dust'. While 'normal' businesses had closed, so too had the 'infamous second-hand store where macabre fridge magnets, T-shirts and mugs were offered as tasteless souvenirs of an event which has stigmatised the town' (Maguire, 2003, 16). At the time, the town was looking at positive ways to promote tourism:

'Some want to exploit what happened but the majority want it to go away,' he said. 'We want to promote the town as a friendly country place, not where bodies were found because the ghouls coming here for a look and to take pictures don't put anything into the local economy. We are working to rebuild the economy.' (Mr Anders cited in Maguire, 2003, 16)

In October 2003, the town celebrated 125 years since it was founded, and the festivities included fetes, fairs and dinners. Despite these positive attempts, the stigma of Snowtown as a rural dystopia remained. Virgin Blue Airlines launched a billboard advertising campaign in 2004 encouraging tourists to 'take advantage of "Bottom of the barrel" prices on flights to Adelaide' (Advertiser, 2004, 5). The campaign ceased quickly after families of the victims complained.

Yet, less than a decade later, Snowtown was again used as an advertisement for the television program *Dexter* with the star, Michael Hall, appearing in an airport waiting for a plane to Adelaide and saying 'Adelaide has more serial killers per capita than any other city in Australia' (cited in Fewster and Innes, 2011, para. 13). At around the same time, *Top Gear* featured Adelaide, calling it 'the home of serial killers':

'South Australia is a pretty but dangerous state. ... Its serial killers make double figures before anyone twigs that someone is missing. ... Being home to serial killers, and sharks, and probably also serial-killing sharks, and heavy rain, Adelaide is thus a grand place to drive away from.' (Ben Smithurst, cited in Fewster and Innes, 2011, paras. 5, 6, 8)

Despite ongoing community strategies to distance themselves from the murders, and to focus tourist attractions elsewhere, international marketing strategies saw a resurgence of murder-themed advertising around the area.

Following these instances of international focus, researchers Sangkyun Kim and Gareth Butler (2015, 82) conducted several site visits to Snowtown and interviewed six 'relatively new' local residents of the community in September and October 2011 about their 'attitudes about the influx of tourists to their area, their reflection on dark tourism popularity, and their views on the benefits and costs of dark tourism'. For many within the community, the increase in tourism was seen as directly benefiting them and their business:

'if it had not happened, Snowtown would be another disappearing country town with nobody coming here. Everyone else made money out of this so why shouldn't the locals of the town? ... Busloads of tourists arrive to see the bank, ever since the murders happened, even more so now. Also, the fridge magnets and other bank related merchandise have become so popular, and even tourists from overseas are now purchasing these items ... often I have had to post them overseas.' (Rose, small business owner cited in Kim and Butler, 2015, 83)

Essentially, for many of the residents, tourism associated with the murders was seen as 'an opportunity to revitalise and rejuvenate the township's economy by capitalising on the potential of dark tourism in Snowtown' (Kim and Butler, 2015, 83). Yet, those that were interviewed did not associate tourism explicitly with 'dark tourism', and as Kim and Butler (2015, 83) state, they 'revealed little understanding of the concept of dark tourism'.

While there was little understanding of the concept of dark tourism, many of the suggestions for further tourism activity reflected key types of dark tourism. For example, community members suggested that the bank vault should be bought and reopened as an educational centre and museum, or that signage should be installed to provide tourists with the 'story of the bank' and to educate people that nearly all of the murders did not occur in Snowtown (Kim and Butler, 2015). As previously explored, several owners of the site have had different agendas, yet, the building keeps being resold privately with no tourism infrastructure suggested.

At the beginning of 2012, the bank was again for sale. This time, the real estate agency listed the building on eBay. Within 24 hours of the listing, eBay had removed the page because the site deemed it 'inappropriate or insensitive to victims of disasters or tragedies' (eBay communication cited in Noonan, 2012, 12). However, listing the property on eBay was not the only novel idea by the real estate agency. Prospective buyers were also required to spend money in Snowtown (and show proof of purchase via a receipt) or donate to the Prostate Cancer Foundation of Australia before they were able to attend the open inspection of the property, of which the agency was expecting hundreds of viewers (Noonan, 2012, 41). The property was bought by a Victorian couple who told the media that tourists continued to visit to take photos. As such, they were considering establishing a memorial centre in the bank part of the property, including installing a 'plaque of respect for the victims' and also as a way to educate tourists and keep them in town longer (Mr Vanderveen cited in Hyde, 2014, para. 14).

The multiple changing of ownership of the bank, coupled with strong desires of the community to distance themselves from notions of dystopia and horror has meant that there is very limited tourism infrastructure in Snowtown to cater to 'dark' tourists. As Coleman (2000) reported, and Kim and Butler (2015, 85) similarly found, many of the residents, particularly the older residents who were born in Snowtown and 'harboured strong emotional ties to the town' were 'strongly against both the media and tourist attention the town had received in recent years'. As such, there has been a conflict between newer residents, where dark tourism was morally acceptable and with older residents who perceived the tourism as inapprehensible from the very beginning (Kim and Butler, 2015). Some of the interviewees in the study argued that tourists were made to feel unwelcome and guilty about being interested in the bank, and as such, quickly left the town after taking pictures, thus losing potential economic income.

An important element in determining the ethics of dark tourism in Snowtown was time. One of the interviewees of Kim and Butler's study made reference to Barry Drew (cited in Coleman's 2000 media story) and how he was 'kicked out of town' because he was selling magnets and 'had set up barrels outside the bank for visitors to take photos inside of them' (Kitty cited in Kim and Butler, 2015, 86). While Barry Drew was engaged in this behaviour in 2000, and it was seen as morally reprehensible, 11 years later in 2011, Kitty felt that some dark tourism activity was acceptable because 'there had been time to heal' (Kim and Butler, 2015, 86).

As the media have framed it: 'this is Snowtown's dilemma: Does it use its dark past to secure a brighter future? Does it remember or forget? Would opening the bank to the public also reopen old wounds?' (Keane and Martin, 2019, para. 16). For Rofe (2013, 268), embracing the possibilities of dark tourism activities would uniquely place Snowtown within the 'Australian rural tourism landscape'. While dark tourism may offer Snowtown an economic boost, in what Kim and Butler (2015, 86) have described as a 'stagnating economy' with 'few opportunities for growth', there is a need to develop sustainable tourism strategies that encourage tourists to travel to and stay in the area overnight.

The transitional nature of tourists has long been recognised, for example, in 2003, Goodfellow (2003, 12) wrote that 'the sightseers only come for their happy snaps and go back to the main highway'. For Kim and Butler:

> It is clear that Snowtown's dark tourism attractions, even if fully promoted, would be insufficient to attract long-term visits, or even overnight stays. Indeed, as previously noted, Snowtown's notoriety was borne from its association with the discovery of the bodies of the murdered victims as opposed to the actual acts of murder. Thus, as Stone (2012) posits dark tourism sites may be incorporated into broader tourism itineraries and Snowtown's location in relation to the Yorke Peninsula and the Clare Valley wine region would perhaps support such an idea. (Kim and Butler, 2015, 87)

Similarly, Rofe (2013, 269) argued that without strong political support and investment in Snowtown, any dark tourism activities created ran the danger of becoming a 'superficial horror-themed visitation site at the lighter end of the dark tourism spectrum'.

It is clear that the community of Snowtown want to attract tourism that focuses on the positive attributes of its locale. There is a need to reclaim the rural idyll from the rural dystopia. The town has invested in sporting fields and murals showcasing the beauty of the area. Yet, there is still recognition that 'the town will always have to contend with its history', with a local resident in 2018 stating that 'she is more convivial about those rolling in to take macabre photographs on the main street, saying at least it "brings people into town"' (Willis, 2018, 28).

## Port Arthur

The site of Port Arthur has already been detailed in this book as a site of dark tourism relating to Australia's convict history. However, Port Arthur has a much more recent tragedy that continues to attract tourism and wider popular culture fascination. While Martin Bryant killed and attacked people outside of the PAHS, the focus here is on the convict heritage site and the impacts it has had on it as a tourism destination. Almost three decades ago, 'between 11 am on Sunday 28 April and 10 am on Monday 29 April 1996, a tragic chapter was added to Port Arthur's history when a lone gunman shot and killed 35 people and wounded 19 others in and around Port Arthur' (PAHSMA, 2009, 31). In what is described as Australia's worst single-shooter mass killing, Martin Bryant killed tourists and workers within the grounds of Port Arthur, including 20 victims within 90 seconds at the site's Broad Arrow Café. The peaceful, idyllic, landscape of Port Arthur was once again reaffirmed as a rural dystopia – a place 'stained by misery, reeking of cruelty, haunted, fearful and cursed' (Tumarkin, 2005, 5 of It Goes On).

The massacre led to the site being closed for a month. After reopening, the site experienced lower levels of tourist attendance and changes in tourist behaviour. Tourists predominantly visited for day trips rather than staying overnight in the three months following this period (Frew, 2012). The period of closure and lower tourist numbers following the event impacted the community life on the Tasman Peninsula (PAHSMA, 2009).

Discussion around how the massacre and, in particular, the café site should be memorialised quickly ensued. The site was partially destroyed eight months after the event 'as an act of mourning' (Mason et al, 2003, 22). Calls to destroy the café fully were made by some staff who wanted the structure demolished because 'gore junkies' were being attracted to the site, and some visitors were asking for 'gruesome details' (Lennon, 2002, 44). However, not all supported the destruction of the site, arguing that having the structure remain offered survivors and the community a place to remember and mourn the victims. The varying community desires led to a range of different outcomes:

Some parts of the site were conserved in accord with each set of values. A new memorial was installed (a cross made of huon pine, initially intended as a temporary marker); the demolition of the café, begun immediately after the tragedy, was halted; and the remaining shell of the building was preserved in a state of stripped-down ruins, cleared of any physical evidence of the shooting, yet clearly marking the actual site as a literal memorial. (Mason et al, 2003, 23)

Eight months after the massacre, and after the trial of Bryant, the café's roof, doors and windows were removed, 'thereby creating a new artificial ruin' that symbolised an 'open wound' and the 'pain that persisted' (Strange, 2000b, 175). The official Port Arthur Memorial Garden was opened four years after the event on 28 April 2000. The four walls of the café remain standing, bare and without a roof. The fireplace and chimney are all that remain inside the building, with an entrance at the front of the building and an exit at the back. A wooden bench has been added in a corner of the ruin to allow space for visitors to sit and mourn/reflect.

Outside the café, a reflective pool was constructed and dominates the Memorial Garden (see Figure 6.2). In one corner of the reflective pool, furthest from the exit of the café is a visual representation of the 35 lives lost: 35 gold leaves of different sizes and in random placement can be seen through the water on a circular plate. As in the café, no interpretive texts or explanations are offered to explain this symbolism to tourists, and Frew (2012, 40) found that 'visitors are often seen to count the leaves to confirm there are 35'.

Otherwise, there are only five other markers with text at the site. The first is the Huon Pine cross, which was originally placed at the waterfront but moved in 2001 to a corner of the Memorial Garden to protect volunteers from passing the cross while conducting tours (Altman, 2006, cited in Frew, 2012) and has the names of those killed listed alphabetically. The second marker dates the opening of the garden and lists the names of the victims. Frew (2012) notes that some victim names are more 'worn' than others on this marker, indicating that visitors are touching these names more frequently than other names listed. The third marker has been placed outside the garden and contains a photograph of the reflective pool with text superimposed over the top outlining the events of the day and the bravery and kindness of the survivors (Frew, 2012). The fourth and fifth markers are located at the reflective pool and offer verses on death and peace.

According to Tumarkin (2005), by 2004 the remains of the café were indistinguishable from the convict ruins, and the memorial gardens blended perfectly into the site. The planting of shrubs at the front of the building helps to obscure the café from plain view, which means that visitors need to make a conscious decision to enter the Memorial Garden (Frew, 2012).

**Figure 6.2:** The Reflective Pool, Port Arthur Historic Site, Tasmania, Australia

Source: AS (nd), '*Monument à Port Arthur*'

Yet, as Mason et al (2003) acknowledge, not all of the steps taken by the Port Arthur Site Management would have been appropriate directly following the event itself. The establishment of a more formal, permanent memorial site, the opening of a new visitor centre in a different location in 1999 and the creation of a brochure focusing on the memorial to help visitors understand the issues would not have been welcomed directly after the event at the time. Three decades later, the site has become as much a part of Australia's national cultural history as the convict aspect of the site. Despite the passage of time, the brochure still informs visitors that they 'prefer not to use the name of the gunman' and that many of the staff at the site do not want to answer questions about the event (PAHSMA, nd, 2).

While the Memorial Gardens have been constructed with quiet reflection and empathy in mind, every tourist will experience such sites in different ways. As Strange (2000a, 27) found, some youths have been seen to 'skylark in the shell of the Broad Arrow Café', demonstrating that 'no mode of interpretation or memorialisation can reliably induce anticipated emotional and intellectual responses'. In addition, this site of serial death will always be linked to the wider convict site. As such, it is hard to differentiate the tourist capacity of the Broad Arrow Café – would tourists still travel to the site if it were not embedded within the broader convict tourist site with impressive tourist infrastructure? Or would the Broad Arrow Café become more like the Belanglo memorial?

# Conclusion

Many of the themes explored within this chapter are not unique to rural Australia, or even to sensational crimes committed by serial or mass murderers. For example, Foote (2003) discusses the geographical landscape of Salem, America, and how there used to be little on-site tourist information about the Salem Witch Trials. Indeed, the place of execution of the victims was not advertised, and local people only knew the general area of the executions (Foote, 2003). Geographer David Lowenthal (1975, 31) observed that 'features recalled with pride are apt to be safeguarded against erosion and vandalism; those that reflect shame may be ignored or expunged from the landscape'. As such, it may be suggested that sites such as Salem, Snowtown, Belanglo Forest and the Broad Arrow Café at Port Arthur are connected more with 'shame' and thus a strategy of ignorance be adopted to expunge the memory from the landscape. Yet, as Foote (2003) recognises, shame is just one element that plays a role in determining what is demolished or forgotten, and many acts of violence and the landscape they occurred on, such as the Nazi atrocities, are successfully turned into monuments and memorials.

There are four ways in which communities respond to landscapes of violence and tragedy:

> *Sanctification* occurs when events are seen to hold some lasting positive meaning that people wish to remember. ... A memorial or monument is the result. *Obliteration* results from particularly shameful events people would prefer to forget – for example, a mass murder or gangster killing. As a consequence all evidence is destroyed or effaced. ... *Designation*, or the marking of a site, simply denotes that something 'important' has happened there. *Rectification* involves removing the signs of violence and tragedy and returning a site to use, implying no lasting positive or negative meaning. (Foote, 2003, 7–8, emphasis added)

Influencing which strategy to adopt lies in 'understanding how people interpreted the events retrospectively as unavoidable accidents, heroic battles, instances of martyrdom, or senseless acts of violence' (Foote, 2003, 5). Similarly, the geographical landscape itself has an important role to play in determining how the site and events are remembered. For example, physical remains, such as the Broad Arrow Café, 'demand interpretation' and reflection (Foote, 2003, 5) and must be demolished to avoid such recognition, whereas the landscape of the Belanglo State Forest had no physical remains of the murders and thus does not necessarily force visitors to reflect on Ivan Milat.

In many cases, communities and governments have clear strategies for dealing with landscapes of violence. Memorials of the Holocaust and sites of war or conflict provide a perfect example of this. Whether such strategies occur quickly after an event will often depend on the type of violence that occurred, the shame attached to the event and whether there is consensus about how to remember or acknowledge the atrocity. According to Foote:

> Salem has never completely resolved how to view the witchcraft scare within its longer history. The question has always been whether to ignore the episode as a brief, shameful anomaly, to recognize it as a valid part of Salem's history, or to honor it as a turning point in American religion. (Foote, 2003, 3)

The result has been a tentative exploration of tourist activities (such as a sound-and-light show telling the story of the witchcraft trials in a small witch museum and a notation of where one victim, Giles Corey, was crushed to death). Leading up to the tercentenary anniversary of the trials, there was heated public debate about erecting a monument to the victims of the trials. Opponents argued that nothing should 'draw public attention to such a shameful event' (Foote, 2003, 5). Despite these objections, a memorial was opened in 1992, indicating some acceptance of the historical significance of the town.

Further research, over a decade later, demonstrates the shift in acceptance of the Salem community to embrace the notoriety of the witch trials. For example, the city logo 'references its witch-related past, a design that could be a sailboat – or a pointed witch hat'; and it hosts a large Halloween-themed festival, Haunted Happenings, which includes witch trials themes (Heidelberg, 2015, 80–81). Throughout October, Haunted Happenings attracts 250,000 (a quarter of the 1 million annual visitors) to Salem, which equates to US $99 million in tourist revenue each year (Heidelberg, 2015, 81).

Within this chapter, each of the sites has similarities for tourism barriers as to those found in Salem before 2005. For example, for Snowtown and Port Arthur, there are aspects of shame attached to both sites – while the shame is different to that of Salem, there is nonetheless shame that such horrific events could occur in such idyllic and close communities. In Port Arthur, calls to destroy the building echo the protests in Salem about erecting a monument that draws attention to shameful events, yet the construction of the memorial in both towns indicates the acceptance of events that have had national significance. The memorial built out of the remains of the Broad Arrow Café effectively recognises it as a valid part of the convict site heritage as well as offering a physical reminder of a turning point in Australia's gun control law debates and reforms. It has become a sanctified site. In comparison, Belanglo State Forest has become a designated site – it has a marker to label the site and remember the victims, yet it has no 'long-term attention and is rarely the focus of regular commemorative rituals' (Foote, 2003, 18).

There are many similarities between the community's response to the 'Snowtown murders' and the community of Plainfield in America concerning Edward Gein's crimes. For example, both rural communities experienced unwanted 'spontaneous' tourism after media exposure of the crimes, and both contemplated demolition of the offending sites to obliterate the 'stain' on the town. In the case of Plainfield, this led to arson destroying the house, while in Snowtown, no such demolition was possible. Instead, there have been repeat attempts to rectify the bank vault in Snowtown with a string of private owners trying to reintegrate the building 'into the activities of everyday life' and thus make it 'innocent' again (Foote, 2003, 19).

The media and advertising play a large role in promoting the existence of such sites and in shaping the public's perception of them. For example, as already noted, the media's early headlines of 'Snowtown Murders' irrevocably shaped the public's perception of the town. Despite local pleas that (all but one of) the murders did not happen in Snowtown, the town's name has become synonymous with murder and depravity. The media exposure also ensured widespread knowledge of the atrocity enabling tourists to travel to the site almost immediately after the event, and over two decades later. Each time the media reports on the case, including the various offenders' bids for parole, the public's attention is returned to the dystopian town. This is not a new phenomenon – similar occurrences of recurrent tourism following media exposure were evident in the Jack the Ripper case. However, with increasingly 'instant' forms of media that are accessible over a much broader geographical area, the potential for unsanctioned dark tourism is high.

Tourism activity itself can force communities to place markers, memorials or tourist infrastructure. In Snowtown, tourism activity and the theft of material from the site led to business owners selling merchandise that eventually saw them 'run out of town' by their fellow townspeople. As Gibson noted:

> Serial murder-related tourism activity is typically disliked and opposed by local residents near the tourist attraction. Proprietors of such venues are accused of making money from the suffering of their neighbors ... such tourism sites are almost always attacked as opportunistic and morbid money-making methods. Because of the typical local disapproval, serial murder-induced tourism is frequently shut down and/or banned. Quite often such sites were demolished. (Gibson, 2006, 567)

Similar to Snowtown locals being judged for making money from dark tourism, the tour group that organised the Belanglo tours was similarly condemned, and tours were, in essence, forcibly shut down. Yet the fact remains that sites of violence and tragedy do attract tourism. In rural Louisiana in the United States, the Bienville Parish erected a marker at Bonnie and Clyde's death site because so many tourists asked where the site was (Foote, 2003). Rurality, it seems, does little to dampen tourist activity and the desire to visit the macabre. Turning such attractions into a successful tourist site, however, is more problematic.

The impact of serial killer and sensational tourism has long-lasting consequences on the local community, as well as the families and friends of the victims. As the case study of Snowtown demonstrates, members of the community were driven out of the town when they wanted to engage with dark tourism activities. Further, residents who relocate to communities like Snowtown become immersed in ongoing debates and challenges. The (many) owners of the Snowtown Bank would have faced pressure not to commercialise the premises, and Kim and Butler (2015) found that while newer residents were more open to dark tourism associated with the bank, the older residents were still resistant, potentially creating conflict between the 'old guard' and the 'new arrivals'.

In the case of the Port Arthur massacre, the families, friends and co-workers of the victims have navigated the tension between memorialising the victims and dealing with the notoriety, against benefiting economically by memorialising the site, acknowledging the victims (while denying the offender any 'fame') and requesting that tourists do not ask questions about the massacre to protect themselves from having to relive the atrocity.

As discussed already, one element for determining when tourism may be appropriate at a site of death and suffering relates to time to heal. As Dale and Robinson (2011, 213) highlight, the 'elapsing of time' and 'the scale and intensity of the event in the human consciousness' affect the suitability of potential dark attractions and/or exhibits. The more recent an event, the 'darker' a site will become as a tourist destination (Stone, 2006). Where locals have had enough time to 'heal', such tourism becomes less problematic (Kim and Butler, 2015, 86). However, the sites within this chapter challenge this assumption. Kim and Butler's research highlighted that 11 years after the bodies were found, Snowtown residents were more open to possible tourist-related activities within the town. Yet, the Belanglo tours were conducted 20 years after Ivan Milat's crimes, and this was not considered sufficient time to enable dark tourist activities in this area.

Drawing on international examples, such as Salem, it is evident that many atrocities do not simply disappear from public interest. While this is problematic for some communities like Snowtown, for others it offers the potential for ongoing economic income, like Port Arthur. It is also possible that tourism strategies will be adopted in the future in places like Snowtown, once the shame has diminished and when more new residents are unconnected to the event.

For rural areas, there is also a danger of a location becoming solely known as a dark tourism destination. As in the case of Snowtown, the locale attracted widespread international attention and several media campaigns have focused on the unsavoury and dystopian nature of the town. There is a danger that if rural places like Snowtown choose to focus marketing campaigns on dark tourist activities they will always be known solely for the 'bodies in the barrels' and turn the community into 'museums of themselves' (Gilbert, 2006, 194). Many places that have experienced historical trauma continue to feel the 'events as vivid and real as today's' events (Lowenthal, 1975, 28). Even the promise of economic prosperity may be insufficient for communities to tie themselves to this fate, and indeed research indicates that tourism can often fail to sustain local economies. Gilbert (2006) found that in many tourist towns employment is dependent upon tourist seasons and can thus be very unpredictable and insecure. Additionally, events outside the control of a tourist site will affect visitor numbers – as Port Arthur experienced after the massacre at Broad Arrow Café and again during the COVID-19 pandemic.

The three case studies offer contrasting approaches to dealing with death and atrocity in rural areas. In Snowtown, there have been attempts at commercialising the area and offering kitsch merchandise (although nowhere near to the same extent as Dahmer or Jack the Ripper tours). While it is not necessarily perceived as 'dark tourism' by the locals, there has been ongoing interest (even from the original bank owners) to capitalise on the murders and help the community financially. In comparison, communities have opted for memorials at Belanglo and Port Arthur, albeit the memorial at Port Arthur has more extensive tourism infrastructure and information available due to its location within an existing dark tourism site. For many people, the memorial site of the massacre stands in contrast to the rest of the site – with the intention that it avoids becoming a 'voyeuristic dark tourism attraction' (Frew, 2012, 46). Despite this, it is impossible to manage tourist experiences, and, as has been noted, some tourists are less than reverential at the site and they may be motivated by voyeuristic intentions.

In the cases of Snowtown and Port Arthur, there were immediate visitors to the site with both sites being visited by 'gore junkies'. In comparison, it does not appear that there was immediate tourist activity at Belanglo following the murders and the more sensational or entertaining forms of tourism were rejected by the community. However, the one commonality between all three areas was the treatment of the killer as monstrous, and the transformation of the beautiful, idyllic rural landscape into one of contamination: a place suitable for monsters to conduct their depraved acts away from the civilising norms of urban life.

Both Ivan Milat and the Snowtown killers lived within the urban landscape, ultimately either killing and/or disposing of their victims in the more isolated rural areas. Yet, despite the attempts by the communities to point out that the murderers lived in urban spaces, society continues to associate these murders with rural landscapes (instead of where the murderers lived and assumed their 'normal' lives). It is in the outback that they became monstrous and committed unthinkable deeds – the rural, in a way, acted as an accomplice to allow these murders to occur and go unnoticed. Consider the vault used to store the murder victims in Snowtown – the vault was initially rented to store kangaroo carcasses. While this seems 'normal' in the 'bush' where hunting is 'acceptable', in contrast, the renting of a space in the civilised urban environment to store dead animals is not 'normal' unless it is associated with a butcher.

In all three areas, the rural Australian landscape, like that of Salem and Plainfield, is now irrevocably associated with perceptions of death, the macabre and monsters. Snowtown has not been able to 'civilise' itself again fully, and all three sites will forever be tainted as being the geographical sites of murder and atrocity. Yet, like Salem and Plainfield, tourists continue their visits to such sites. As Smith outlined:

> Snowtown, Belanglo State Forest, Port Arthur. They are places marked forever by the depraved acts of Australia's most disturbed criminals. Strangely, they are also marked on maps as holiday destinations to tour, where adventure awaits and where one can walk in the footsteps of a mass murderer. (Smith, 2015, paras. 2–3)

No urban environment is similarly marked so uniquely and sweepingly. Furthermore, where urban environments have been sites of atrocity, the atrocity is absorbed by the location and 'hidden' by other attractions and events. Communities can embark on dark tourism activities, but there is not the same economic pressure to engage in activities that are disrespectful to the communities. Perhaps this is why the Belanglo Tours were rejected so effectively – Goulburn, being a much larger regional town, had the ability to specify what was acceptable. The uncivilising and remote nature of the Australian bush and rural landscape ensure that sites of death and atrocity remain synonymous with danger and continue to attract tourists.

7

# Ghost and Crime Tours, Deadly Towns and Disaster Tourism

This chapter looks at 'deadly towns', ghost and crime tours, tourism associated with natural disasters and the growing nature of online dark tourism, which may bolster knowledge of rural locations too remote to access in person. According to Addeo et al (2021a, 280), the role of the media has increased dark and disaster tourism, and 'once a disaster has been publicised, it becomes an attraction to those who, for whatever reason, wish to travel to gaze upon it' (Sharpley and Wright, 2018, 336). Further, many of the locations discussed in this chapter are not formally recognised tourist destinations and some have an associated level of danger attached to visiting these sites. For certain tourists, there may be excitement, pleasure and heightened emotions linked with the prospect of visiting a site where there are restrictions and prohibitions (Addeo et al, 2021b, 213).

Furthermore, the increased use of social media is both encouraging tourism and offering new avenues for 'dark tourism'. For example, as will be explored in this chapter, Instagram has been accused of motivating tourists to travel to remote and risky destinations in search of the perfect selfie; and online social media sites offer free advertising as well as 'shared spaces online for dark tourists to communicate, share experiences and discuss how such places affect and influence them' (Bolan and Simone-Charteris, 2018, 731). There are also dedicated 'dark tourism' websites directing people to various locations across the globe, and a growing number of publications focus on advertising such destinations. For example, Peter Hohenhaus (2021) recently published *Atlas of Dark Destinations: Explore the World of Dark Tourism* advertising (and rating by 'darkness') dark sites across the world.

## Ghost and crime tourism

Notoriety can often overtake small towns. Impromptu tourism to sites of murder and crime continues to occur across the world, and the sites themselves become irrevocably intertwined with the name of the location. As Heidelberg (2015, 74) outlines, following the killing of the DeFeo family by their son Ronald on 13 November 1974, Amityville, United States, 'is no longer a quiet, small town community; it is linked to the Amityville Horror and is a dark tourism destination'. Many communities have scandals or tragedies, but in many rural locations, these ghosts become the new town identity, and 'give small communities like Amityville a level of long-term fame they never anticipated. Perhaps this fame is not something the community sought or wanted, but choice is ripped away with widespread media attention' (Heidelberg, 2015, 75).

As seen in the case of Snowtown, trying to silence tourism does not work. Conflict exists within these communities, where residents clash over whether to capitalise on the horror or try to quash tourism. Ignoring the tourism potential limits the narrative that can be shaped around the event and the impacts on the town itself (Heidelberg, 2015). Other towns actively embrace dark tourism:

> Owners of the Lizzie Borden house in Fall River, Massachusetts, operate the site as a bed and breakfast. The Villisca, Iowa axe murder house hosts tours. For a price, visitors can stay overnight in the 'haunted' house. The house's story and haunted reputation have been featured on paranormal investigation shows such as Ghost Adventures, attracting visitors who otherwise might have never known of Villisca's existence. (Heidelberg, 2015, 82–83)

Ghost and crime tours are another common form of 'dark tourism' within Australia; and while many (if not most) occur within major cities, there are several rural areas that use ghost and/or crime tours as a marketing strategy. For example, Kapunda in South Australia advertise that they are 'not only the most haunted town in Australia, but home to the most haunted Hotel in Australia' (Ghost Crime Tours, 2019a, para. 1), and an Adelaide-based company runs ghost crime tours where they explore the more grisly parts of the town's history. Some have suggested that Kapunda is the most haunted town in the world (Ryan So Fly, 2021).

Kapunda has a population of 2,917 and is advertised as the 'rural centre for the Mid-North' of South Australia (Light Regional Council, 2020, para. 2). While the local council focuses on advertising arts, culture and the history of the town, an image of the North Kapunda Hotel ('the most haunted Hotel in Australia') is featured, thus directing tourists to this 'dark' site. Despite the lack of overt advertising by the local council, the tour and hotel are viewed as key tourism destinations within South Australia. For example, the tour was featured in 2014 on the TV series *Haunting: Australia* and Virgin Australia highlighted the town in their online article 'A haunted tour of South Australia, 8 macabre experiences' (indeed, it made the top of the list above the Adelaide Gaol).

In 2001, the town became famous after ghost hunters filmed the area. As such, Kapunda became notorious for ghosts, and rather than shying away from the reputation, an Adelaide-based company decided to capitalise on this and started running tours. However, the tours did not incorporate all of the 'dark' sites within the town limits, and unsanctioned tourism led to the demolition of the St John's Church Reformatory (located 10 minutes out of Kapunda) in 2002, due to safety concerns with tourists entering an unsafe building (Ryan So Fly, 2021). The Reformatory originally housed young females in 1897, and the building included 'holding cells' for those girls who tried to escape. Today, tourists continue to visit the cemetery, which is adjacent to the demolished reformatory.

As part of the official tour, participants are taken on a two-hour walking Ghost Crime Tour of Kapunda, where tourists are provided with a 'history tour of murder, suicide and hauntings from the 1800s' (Lewis, 2017, para. 1); providing 'details of a crime whilst standing at the exact location where it took place' (Tripadvisor, 2022, para. 1). There are meant to be several places on the tour where participants can encounter ghosts. However, the focus is on the hotel where several deaths occurred in the sealed accommodation rooms upstairs, which have been left unattended: 'the crumbling hallway and rooms are a horror movie in waiting' (Lewis, 2017, para. 4). For some, the walking tour fell short of a satisfactory experience. Rod Lewis provided a less than amorous review in 2017, largely attributing the disappointing tour to the inexperienced guide leading the tour. However, he also noted the lack of in-depth detail of Kapunda's historical (and Aboriginal) past, and the overwhelming focus on the hotel, rather than on the town as advertised. This may be a result of the company running the tours being based in Adelaide, rather than being organised at a grass-roots level.

Tourists could even opt to participate in a sleepover at the North Kapunda Hotel where 'several people were murdered ... including ladies of the night' (Virgin Australia, 2016, para. 2). The sleepover has now been replaced with a Paranormal Lockin, where the focus is on paranormal activity, rather than crime stories. Tourists are advised to take the ghost crime tour before participating in the Paranormal Lockin to understand the history and stories of the hotel (Ghost Crime Tours, 2019b). Reinforcing the focus on the paranormal, people are now able to log in via the Ghost Crime Tours Website to 'live stream' videos (login required) from the main areas of the North Kapunda Hotel: the 'Hallway to Hell', 'Sarah's Room', the cellar and the servants' quarters.

There are few references to the rurality of this location. Even in online videos where Kapunda is described as being 86 kilometres away from Adelaide, or a one-hour-and-10-minute drive, it is referenced just as an 'interesting little town' (Ryan So Fly, 2021). The scenery along the drive, and at the nearby St John's Church Reformatory, highlight the geographical isolation of the town, yet this is not discussed. The only allusion to the geographical isolation of the area is in reference to swallowing flies while in the 'country' (Ryan So Fly, 2021).

## Deadly towns

There are an increasing number of 'deadly towns' across the globe, many attracting 'dark tourists'. For example, Chernobyl in Ukraine has long been classified as a popular 'dark tourism' site where 'tourists are excited to wear protective clothes, to grab their own dosimeter to measure radioactivity', with 'no thoughts' to the health risks (Marton et al, 2020, 144). On 26 April 1986, the world's worst nuclear accident occurred at the Chernobyl nuclear power plant in the Ukrainian Soviet Socialist Republic (now Ukraine). Despite ongoing health and safety concerns with the radioactive 'fallout' of the site, 'illegal tourism to Chernobyl has flourished', to the extent that the Ukrainian government sanctioned official tours to the site, as well as to the nearby abandoned 'ghost town' of Pripyat in 2011 during the 25th anniversary of the disaster (Stone, 2013, 79). Chernobyl offers tourists the ability to witness the transgression of the boundaries of normalcy that have been frozen in time in an area 'not freely accessible' in a post-apocalyptic place (Stone, 2013, 81).

Within the Chernobyl tour zone, tourists must be over 18 years of age and wear clothes that cover their body (shorts and open footwear are forbidden). Tourists are prohibited from eating, drinking or smoking within the zone, touching buildings and trees, sitting or even putting personal belongings on the ground (Chernobyl Tour, 2021a). Tourists are also warned of the ethical aspects of the site and encouraged to behave accordingly. As part of this, taking pictures or videos is only allowed when the guide gives permission; visitors are asked to be mindful of what they say and how they behave on the tour; and no objects may be taken from the site (Chernobyl Tour, 2021a). Tourists can, however, take home purchased souvenirs of their tour, including various 'airs' of Chernobyl (air trapped in a can, which the site promotes as 'Safely canned Air from the Exclusion Zone is a must have for radiophiles and a cool gift for radiophobes' (Chernobyl Tour, 2021b, para. 7)) and many other glowing ('radioactive') themed shirts, mugs, maps, hats, pens, magnets, keychains and lighters.

In Australia, the town of Wittenoom has a similar 'dark' history. Wittenoom is located 1,425 kilometres (880 miles) north of Perth. The area was mined for blue asbestos from the 1930s until 1966 when the mines were closed over health concerns. The town's population has always been low, with the highest non-Aboriginal population recorded in 1961 with 881 residents

(ABS 1962)[1], and it is estimated that more than 20,000 people have lived in Wittenoom over the time the mine was functioning. However, there are many more Aboriginal Australians affected, as well as tourists and visitors to the town over the years.

In 1978, the government of Western Australia 'made the unprecedented decision to phase down the town' (Western Australia, Legislative Council, 2021, para. 3). As a part of this process, the status of Wittenoom as a township was revoked in 2007. On 24 March 2022, the Western Australian parliament passed the Wittenoom Closure Bill which enabled the compulsory acquisition of privately-owned land in the area and provided the necessary regulations to allow the government to demolish all above-ground infrastructure 'within the former town site to limit the attraction and opportunity for people to visit and stay in the area' (Western Australia, Legislative Council, 2021, para. 4). By the end of September 2022, all residents had been removed from Wittenoom.

The town has been labelled as the most dangerous and deadly in Australia because over 2,000 people have died from asbestos-related diseases (Mining Editor, 2014), and it is estimated that 25 per cent of all men who worked in the mines will die from similar causes (Aussie Towns, 2021). Further, the families of workers, travellers and of course Aboriginal people were also exposed, as much of the town's infrastructure was also contaminated. For example, the town swimming pool had a fake beach created using asbestos tailing (waste from the mill); the sand pits for children were similarly made of the same material, as were the surfacing of the streets, footpaths, parking areas, local racecourse and school playgrounds (Mining Editor, 2014). Similar material was spread over the yards of houses to suppress red dust and mud (Reid et al, 2007). Trucks carrying asbestos travelled past the local primary school and residential areas, and miners wore their contaminated clothes home, infecting the entire household (Reid et al, 2007).

In addition, the landscape has been irreparably damaged, with long-term consequences for the land and its traditional owners (the land was returned to the Banjima people in 2014), many of whom are also suffering from asbestos-related diseases. The land was never cleaned and stands as a rural dystopia within the natural beauty of the landscape:

> The blue-grey mounds of the old Colonial and Wittenoom mines stand out against the red ridgelines and green valleys of scenic Wittenoom Gorge. The riverbeds and creeks are the colour of asbestos, where the tailings have washed down over decades of annual floods, past

[1]   Prior to 1971, the Australian census only counted non-Aboriginal people, as such, the population was higher in 1961.

the condemned townsite, through tributaries, towards the mighty Fortescue River. (Bennett, 2015, para. 10)

The contaminated site comprises 50,000 hectares (120,000 acres), and blue asbestos fibres remain on the ground, in the air and waterways, including the Fortescue River catchment, which will be impacted for hundreds of years if the area is not cleaned (Bennett, 2015).

Similar to Chernobyl, tourists began to visit Wittenoom when the government was trying to close the area in 2007, and the town has been dubbed #Australia'sChernobyl on Instagram (Carroll, 2020). Journalist Erin Lyons states that Wittenoom is:

known as one of Australia's most deadly towns and considered the most contaminated site in the Southern Hemisphere. Despite the danger, it has become a site for dark tourism, with social media influencers and thrill-seekers eager to have a look at the erased community. (Lyons, 2020, paras. 18–19)

In *Fodors Travel*, Wittenoom made the top ten tourist destinations (the fifth) to see in Western Australia (O'Connell, 2020). According to the article:

The rising popularity of dark tourism, where travelers visit eerie destinations, has seen the tiny town of Wittenoom become one of the state's most unusual and dangerous attractions. ... Health authorities warn that even visitors to Wittenoom are at risk from the asbestos fibers in the air, but it hasn't stopped tourists from going to this ghost town to pose for photos. (O'Connell, 2020, para. 8)

The town of Wittenoom provides tourists with an opportunity to visit a 'deadly' 'ghost' site, which continues to pose health risks to those who travel there.

Given the very few residents living in the area in the last decade that Wittenoom was a 'town', the prospect of tourism would not have increased the economic prosperity of the town. Additionally, the location of Wittenoom highlights the extent to which tourists will travel to rural and remote areas to witness dark periods of history. Tourists to Wittenoom explore the area, including the natural landscape, the mines and also homes within the town where 'entire homes filled with household items and furniture now gather dust after residents were forced to flee and leave their belongings behind' (Scott, 2019, para. 7).

Although the exact numbers of tourists are unknown, one media article reported that 60 tourists a week 'risked their lives to visit the "ghost town"' (Towie, 2022, para. 9). Other anecdotal evidence suggests that people hire four-wheel drives to tick the site off their 'bucket list' (Bodie Norman cited in Birch and Gorman, 2021). While many tourists try to wear protective equipment (P2 mask or PPE), others do not. One self-proclaimed tourist to the area, Mitchell Shand, camped overnight at Wittenoom because he was 'curious about the forbidden town', and while he wore safety PPE outside his car, he was still worried about possible health implications (Birch and Gorman, 2021, para. 7).

Yet, the vast majority of images on social media relating to Wittenoom show visitors without proper protective gear. For example, one photo featured in the media was of a tourist wearing undies as a facemask to 'protect' against breathing in any hazardous material (Zaunmayr, 2019). Images show people taking selfies in front of danger signs or at the entrance of the former blue asbestos mine without protective wear (Carroll, 2020). The recognised danger of the area, and the levity with which it is often treated, is evident in the 'souvenirs' that were previously on sale. While there were residents living in the town, tourists could visit the local gem and souvenir shop and buy car stickers and magnets reading 'I've been to Wittenoom and Lived' (Dunn, 2008, para. 1).

The media have been quick to focus on the 'dangerous' and 'extreme' nature of tourism to Wittenoom. Particularly, the media has disseminated images of tourists ignoring safety warnings and searching for artefacts to take home as souvenirs. For example, De Poloni (2018, para. 9) reported that guided tours were being offered to the area, that people were taking their babies to the town, and that the site is 'an intrigue that lends itself to a form of "extreme tourism" appealing to more thrill-seeking travellers'. Tourists were also reported as defacing and vandalising the warning signs and taking items from the area as 'hunting trophies' (Lyniece Bolitho cited in De Poloni, 2018, para. 29). Other articles reported that 'Insta-fame hungry tourists … [were] ignoring cancer warnings to chase trophy photos' and showed tourists 'with no protective gear, sometimes just a shirt covering their mouths, walking on asbestos piles and even breaking into dust-filled mine shafts' (Zaunmayr, 2019, paras. 1–2).

Images online support these statements, with people often posing in front of the warning signs and even signs that have been defaced. For example, one Facebook photo shows a man standing in front of a 'caution' sign graffitied so that the message 'this area is subject to flash flooding' reads 'this area is subject to scary zombies' with a tag of Whittnoom [sic]. Other photos captured tourists standing in front of warning signs and holding 'trophies' of riebeckite (which is the fibrous form of one of the types of asbestos mined in Wittenoom) (Carroll, 2020).

Given the ongoing life-threatening danger posed to tourists visiting Wittenoom, various strategies have been suggested to deter tourists. For example, numerous warning signs (see Figure 7.1 as an example) seem to do little to deter tourists. Consequently, there have been calls for the roads leading into the town to be cut off; however, as the president of one of the four-wheel drive clubs outlined, erecting barriers may just increase the enticement to the 'mystery', 'Saying, "Danger, do not enter", it's like having a bag of lollies in the cupboard and saying, "Kids, don't eat these lollies". People are still going to do it' (Bodie Norman cited in Birch and Gorman, 2021, para. 27). As noted, one of the strategies to decrease tourism has been to relocate local residents and demolish the town buildings. With the passing of the Wittenoom Closure Bill, the media were quick to publish articles with headlines 'Last remaining Wittenoom properties to be demolished in bid to deter danger-seeking tourists' (Brookes, 2022). However, others believe that the government's desire to close and demolish the town will attract 'a new wave of tourists to the site' by increasing its notoriety and the desire to take 'eerie travel shots' for Instagram (Scott, 2019, para. 9).

In the reporting, while it is acknowledged that the township has seen death and suffering, there is no discussion about how 'dark' this is, or the possible impact on families of those who have died as a result of the asbestos mining or those who have an asbestos-related disease and are suffering. One media article spoke to a former Wittenoom 'child' (now an adult who was diagnosed with an asbestos-related disease in 2019) about the tourism,

**Figure 7.1:** Warning sign for Wittenoom, Western Australia

Note: While the sign is designed to discourage tourists, some have suggested that it is more likely to attract tourists.

Source: Alan (nd), 'Danger Sign in Wittenoom, Pilbara, Western Australia – the town not on any map due [to] asbestos problems'

their response, '"Don't go. You are stupid, you are mad, you are insane," she said. "Asbestos is deadly, scary, microscopic s★★★, and it will get you"' (Helen Cheeseman cited in Zaunmayr, 2019, para. 4). The focus within the media remains on the safety of the tourists ignoring warning signs, rather than any ethical questions over tourists visiting a site that has killed at least 2,000 people.

However, not all tourists are 'dark' or 'extreme' tourists. Despite the dire headlines about extreme tourism, there is recognition that many tourists travel to the area for the natural beauty – the gorges and the wildflowers. The town and its surroundings have been described as an 'eerie beauty' (De Poloni, 2018, para. 9), and it is surrounded by other natural landscapes that attract tourists and campers.

There are also discussions to create a memorial to the victims of the mines, which has the potential to attract tourism to a non-lethal site. The Asbestos Diseases Society of Australia (ADSA) created an online petition to create support for a permanent memorial to be erected in Karijini National Park. The memorial will serve a dual purpose: first to memorialise through featuring the names of the thousands of Wittenoom workers, residents, traditional owners and their family members who lost their lives to asbestos-related diseases, and second, to act as a deterrent to tourists considering risking the area (Newton, 2021). When the memorial is constructed, space will be left for the inevitable victims that continue to emerge from Wittenoom, including those tourists who have not protected themselves. For example, the Chief Executive Officer of ADSA, Melita Markey, has stated, 'Sadly, we know the death toll from Wittenoom won't stop, until the visitors stop' (cited in The West Australian, 2021, para. 2).

One potential 'solution', though likely unpopular for the government, would be to establish formal tours similar to the Chernobyl tours. Tour companies could establish (under regulation) a 'zone' and control the types of tourists entering, and ensure protective clothing is worn and so forth. This would stop families from taking children in (if the 18 years and older rule is to be applied), and ensure that tourism continues in a 'safe(r)' environment. However, given the region has been returned to the Traditional Owners, tourism may further destroy the land, and as such, any possible tourism endeavours would need to be run through or in conjunction with the Aboriginal community.

## Natural catastrophe and post-disaster tourism

There is a growing 'niche segment' of tourists that are travelling to 'dark' sites where natural disasters and catastrophes have occurred (Marton et al, 2020, 137). For example, tourists worldwide travel to witness post-disaster scenes of earthquakes, hurricanes, bushfires and floods. Wright and Sharpley

(2018) researched tourism to L'Aquila, the capital of the Abruzzo region in central Italy, where an earthquake destroyed much of the city's historic centre on 6 April 2009. Furthering the 'darkness' of the site, and as Wright and Sharpley (2018) argue, making this disaster a tourism attraction, 308 people lost their lives, 1,500 were injured and the majority of the population was made homeless due to the destruction caused. Tourists arrived within days to take photographs of the destroyed city (Wright and Sharpley, 2018).

The loss of human life in particular transforms disaster tourism into a category of dark tourism, and this form of tourism is often a temporary phenomenon until the site has been rebuilt. However, some sites, such as Pompeii, become permanent disaster dark tourism sites. A further distinguishing feature of disaster dark tourism is that visitation is 'relatively immediate, spontaneous and unmanaged' (Wright and Sharpley, 2018, 164). More formalised tourism at disaster sites offers governments or businesses the opportunity to reframe the narrative into one of 'hope, survival and renewal' (Miller cited in Tucker et al, 2017, 309).

It is important to note that there is a variety of reasons why someone might travel to a post-disaster location: they may want to aid in the recovery of a location; have a personal connection to the place that they need to check on or try to help protect/restore (family, friends, property, spiritual connection and so on); or they may be curious. A Hungarian study found that most respondents (sample size of 206) would not travel to a site of a natural catastrophe (87 per cent), with 73 per cent actively avoiding such destinations (Marton et al, 2020). Those people more likely to travel to sites of catastrophe, or death, were within the younger generation of 18–25 years of age.

As Marton et al (2020, 137) note, a 'catastrophe tourist does not equal to a deviant person', yet, as we saw in Wittenoom, it may be that such tourists may engage in risky behaviour to visit a location. As such, while the motivation may be for adventure, exploration, mystery – or, as many journalists were quick to point out in regard to Wittenoom, in search of the best Instagram selfie – tourists need to consider the risk to themselves and others in the community. In addition, immediate tourism to an area may be seen as disrespectful and anger local communities (Wright and Sharpley, 2018). For example, following the Christchurch (New Zealand) earthquake devastation in 2010, residents were angry with tourists, whom they perceived as 'rubber-necking' or 'rubble-necking' (Tucker et al, 2017, 315). Christchurch blog writer, Margaret Agnew, described the activity of 'rubble-necking' as those 'tourists who like to drive through disaster zones gawping' (cited in Tucker et al, 2017, 315).

Despite this, research has highlighted that there are some benefits to visiting dark tourism sites of natural disasters. For example, Zhang (2022, 5) argued that visiting such sites enabled visitors to 'depersonalise' themselves as individuals and consider themselves as 'human' or, in other words, to see themselves as 'being small and fragile, being resilient and great, being benevolent, being destructive, and having uncertain fate'. At the same time, visitors can understand human resilience and the ability of individuals and communities to overcome disasters. It can also make visitors appreciate their own lives more, to reconnect to their life (Wright and Sharpley, 2018).

In certain types of disaster sites, the 'dark' side of humanity is also revealed, whether it be destruction to the environment or the industrial neglect of the welfare of individuals. Fostering a sense of human identity at tourism sites dealing with post-disaster sites will inevitably move a site along the 'darker' side of the dark tourism spectrum – where tourists are encouraged to empathise and confront mortality. Yet this will also increase authenticity where sufficient representation and interpretative resources are integrated (Zhang, 2022). Longer-term tourism also offers opportunities for remembrance, which can be very important for communities. Through tourism, the memory of the disaster can be 'kept alive' (Tucker et al, 2017, 310).

## Bushfires

Bushfires are a reoccurring natural disaster in many parts of the world, and the news often features graphic images of bushfire destruction and death across the globe. Reports focus on the loss of lives (both human and animal) as well as property damage. Bushfires cause intense devastation, chaos and death within (mainly) rural and regional communities.

The Australian spring/summer of 2019–2020, also called the 'Black Summer' bushfires:

> saw a period of intense, high-severity fires spread across Australia, burning an immense 17 million hectares of land. Thirty-three people, including nine firefighters, lost their lives, approximately 3,000 homes were destroyed, and over one billion animals were killed or injured. (Heritage Manager, 2021, para. 1)

Intense media coverage showcased the devastation caused by the bushfires to the nation and to the world.

Before the COVID-19 pandemic, the bushfires dominated the media on an international level. In January 2020, journalist Erin Lyons predicted:

> Apocalyptic scenes of charred land, dead animals and skies blanketed in smoke are not the typical visuals tourists associate with a holiday in Australia – but they could encourage a new type of traveller to look our way ... could the destruction and devastation attract a different

wave of tourism – dark tourists – to the impacted areas? (Lyons, 2020, paras. 1, 5)

Within the article, experts were quick to point out that tourists to bushfire-affected sites should not be called 'dark tourists' because their 'motivations are totally different' (Gabby Walters cited in Lyons, 2020, para. 10). Rather than wanting to see the disaster, local and national visitors would be more likely to want to visit to help and support the community. Specifically, 'no Australian wants to be seen as a rubberneck, as a ghoulish character turning up to these place [sic] to have sneaky look' (Lyons, 2020, para. 15). International visitors, however, were tarnished as 'dark tourists' or 'sensation seekers' with the belief that they would be likely to take selfies in front of the devastation because they are not 'emotionally connected' to the tragedy (Gabby Walters cited in Lyons, 2020, para. 14).

Communities affected most by bushfires can be placed in similar situations to those tarnished by serial killers, where tourism may be necessary to restore economic prosperity to the location. Terry Robertson, the Chief Executive Officer of Destination Gippsland (an area located in Victoria's south-east) recognised that any form of tourism, dark included, would be welcomed because it was seen as a form of monetary assistance:

We need to be respectful. We won't discourage or encourage (dark tourism) but if it's handled appropriately visitation to those areas is welcome as long as it's done in a way that doesn't harm the environment, and isn't unsafe. (cited in Lyons, 2020, para. 27)

Tourism to locations affected by bushfires, and fears of dark tourism activity, became overshadowed by COVID-19 and lockdowns. The Australian border was rapidly closed to international tourists, and even local areas were quickly locked down, meaning that 'tourism' was almost non-existent for two years across Australia, with Victoria experiencing the most lockdowns. As such, by the time tourism began to return, many of the areas had regenerated to some extent, thus limiting 'dark' or 'sensation' seekers.

Despite this, other forms of tourism relating to bushfires were promoted. In 2020, the media announced that the National Museum of Australia in Canberra was collecting images, videos and objects from across Australia to display in a *Black Summer* exhibit. One of the featured artefacts is a melted phone booth from Cobargo, which curator Libby Stewart described as 'a very powerful object with the plastic all melted down – it looks like a dystopian painting' (cited in Wheaton and Reardon, 2020, para. 4). Curators intended on spending at least a year visiting bushfire-affected communities across Australia to collect artefacts for the exhibit. The intended exhibit was welcomed by those rural and regional communities affected by bushfires because they wanted people to remember the loss and destruction caused by the fires. Some residents believed that the COVID-19 pandemic had overshadowed the victims of bushfires, limiting the empathy, support and understanding of these victims (Wheaton and Reardon, 2020).

The Cobargo melted phone booth is one of the many exhibits on display at the Great Southern Land exhibition at the National Museum of Australia, and an online short video has been made available to the public, presented by one of the local residents of Cobargo. Rhonda Ayliffe talked about the symbolism of the melted phone booth and:

> how major infrastructure fails, and community doesn't. When you live in a small community, you know everyone, and everyone knows you. So, in the immediate aftermath, people looked after people care ... the very best things that any small place is, is community. That's the heart of any place. (cited in National Museum of Australia, 2022, Cobargo phone box video)

While the focus is certainly not on the geographical remoteness of Cobargo, Rhonda's speech about 'small communities' and 'small places' is one of the few references to rurality within the exhibit. The focus is very much on the devastation and needing to remember and learn, rather than geographical variations.

Some rural locations have taken steps to exhibit their own experiences of bushfires in the local area. For example, a Black Summer and Beyond exhibition of images from the Macleay region was exhibited in the small rural town of Willawarrin in 2020, and the exhibit moved across the region to two other towns to enable multiple locations to commemorate the devastation (Pascoe, 2020). Such exhibits empower local people to tell their own stories and reflect on their past. It also offers tourists to the area an insight into what befell the town.

## Digital dark tourism

As explored throughout this book, most dark tourism sites host their own web pages and promote their tours via social media. As Bolan and Simone-Charteris argue, dark tourism is a growing phenomenon on social media:

> On the more 'entertainment' side of dark tourism, the London-based 'Jack the Ripper Tour', for instance, has developed a sizeable following and growing presence on Twitter (with over 1300 followers). Jim Morrison's grave in Paris has its own Facebook page, whilst on the darker side of Stone's (2006) taxonomical 'dark tourism spectrum', the former German Nazi concentration camp at Auschwitz-Birkenau in

Poland has a number of highly followed social media pages on various platforms including Facebook, Instagram and Pinterest. (Bolan and Simone-Charteris, 2018, 730)

These sites offer visitors the opportunity to share their own experiences with other tourists and allow prospective tourists glimpses of what the site entails. Importantly, tourists use social media sites to:

> mark their visit there, and to comment on what they saw and what they felt, and what they experienced. Furthermore, they want to not only share this but also to participate (at least in a virtual sense) with others who have had similar feelings and experiences. (Bolan and Simone-Charteris, 2018, 731)

As such, social media provides dark tourists with an online community that fosters 'social interaction, communication and information dissemination' about their personal experiences and how the site affected them (Bolan and Simone-Charteris, 2018, 731).

With the pandemic blocking tourism for so long, the digital imprint of dark tourist sites has increased, and tourists can take 'virtual tours' online. However, as Bolan and Simone-Charteris argue, online offerings can never replace a 'physical visit' to a site, because:

> For example, to fully participate in the convict experience at sites of former prisons, visitors will have to feel the cold dampness of a stone prison wall, feel the claustrophobic conditions of a cramped cell, lie on an uncomfortable prison mattress and feel every lump and bump of the basic metal bed. (Bolan and Simone-Charteris, 2018, 732)

As such, digital tourism offers an extra layer of marketing. However, it may be the case that rural sites may be accessed more frequently online than they are potentially in person due to the difficulty of travel (particularly through lockdowns and the like). Sites need to be careful to attract the interest of visitors to make them want to travel long distances, while not giving too much away that enables tourists to just visit 'online' and thus minimise their economic return.

Non-physical tourism offers tourists several advantages. First, there is an illusion of authenticity (Addeo et al, 2021a), particularly if the digital experience is interactive (perhaps a 360-degree video, or the ability for people to click on items to garner further information about a site/object). Second, it can allow tourists into an area in the privacy of their own homes, meaning that their reactions can be uncensored. However, this also means that the true emotional connection with a site is likely to be lost or overlooked. Sites may focus more on 'entertainment' to cater to engaging viewers who are likely to spend a limited time online and consequently may lose sensitivity to the issues being presented (Bolan and Simone-Charteris, 2018). As such, there may be ethical issues where empathy or human connection is lost in an online environment. Third, constant 'livestreaming' options, like at the North Kapunda Hotel, offer tourists an ongoing opportunity to experience something 'dark' that perhaps they missed during a physical tour (or perhaps they were not able to travel to the actual tour, and thus the virtual experience provides a more accessible option).

There are 'dark tourist' websites (and now travel books) that direct tourists to sites worldwide. Over the last decade, there have been a growing number of mobile applications that encourage tourism to dark places and provide an additional level of information, entertainment and 'gaming', particularly at sites that may have little to no tourism infrastructure. For example, one:

> mobile app entitled 'Killer GPS: Crime Scene, Murder Locations and Serial Killers' which provides users with over 470 locations around the globe where infamous murders were committed or bodies were dumped. Interactive maps, travel directions and profiles of famous serial killers are all provided through the app to the user's phone or tablet. (Bolan and Simone-Charteris, 2018, 736)

Similar to the Ned Kelly Touring Route, tourists are encouraged to travel to different geographical areas to experience a themescape. While Australia does not yet offer similar large geographically orientated apps, there are certainly apps that guide tourists around 'dark' sites in the city.

## Conclusion

Disaster, 'ghost' and 'deadly town' tourism allow tourists an opportunity to explore ontological insecurities, existential anxieties and social identities. As Zhang has written:

> Climate change, a global pandemic (COVID-19), bushfires and wildfires in Australia and California will see this era remembered as one of unprecedented human crises, making individuals existentially uncertain and ontologically insecure. (Zhang, 2022, 1)

Tourists visiting these sites are likely to rethink their everyday life, self-identity, family identity and human identity (Zhang, 2022). As Lennon and Foley (2000) originally outlined, dark tourism offers visitors a platform from which to view death and destruction, to confront anxieties surrounding mortality and the 'evil' of atrocities.

Given the range of possible tourist sites covered in this chapter, it is unsurprising that several of Phillip Stone's typologies apply here. For example, the bushfire exhibit at the National Museum of Australia can be categorised as a dark exhibition which offers products that 'revolve around death, suffering or the macabre with an often commemorative, educational and reflective message' (Stone, 2006, 153). As the 'exhibits' are authentic and represent the death and devastation of human life, animals, landscapes and homes, such exhibits are on the darker periphery of the 'dark tourism spectrum', even though they have been relocated away from the actual site of destruction. Despite this, the artefacts retain their macabre 'feel', effectively transporting tourists to memories of natural disasters (either 'real' memories for those who have experienced it first-hand, or perhaps memories of media coverage at the time). Yet, these exhibits are not on display to titillate consumers, rather they are there to serve as a reminder of the power of climate change and our natural landscape, to commemorate those who lost their lives, loved ones or property, and to educate future generations (or those who have not experienced such natural disasters) about strategies to minimise such tragedies. Or, as Zhang (2022, 7) outlines, such exhibits 'demonstrate people's achievements and their capacity to learn from disasters, adapt and carry on, which contributes to a sense of human identity'. The focus on disaster recovery can also lessen the 'darkness' of such exhibits, by instilling hope in visitors.

The cemetery at St John's Church Reformatory provides tourists with a 'Dark Resting Place', a label suggested by Seaton (2002, cited in Stone 2006) to describe dark tourism travel to cemeteries or grave markers. However, the lack of organised tourism infrastructure for the cemetery means a 'darker' atmosphere and ensures that tourists must travel by themselves. The walking ghost and crime tour of Kapunda could be categorised as a dark fun factory, with a focus on entertainment, presenting macabre deaths (rather than general history) and ghost hunting. At the same time, the tour (and the Paranormal Lockin) certainly is not family-friendly, so while they remain 'fun-centric', they would still remain on the darker side of the spectrum.

Touring disaster sites or deadly towns fits within the classification 'Dark Camps of Genocide', where 'atrocity and catastrophe' are the main thanatological themes. However, in the case of Wittenoom, because it lacks tourism infrastructure similar to Chernobyl, which forces tourists to reflect on the darkness of the site, the location is often treated 'lightly' by tourists who ignore warning signs in the pursuit of risk and Instagram images.

Given the death and destruction covered throughout this chapter – death from disasters, asbestos and murders – it is understandable to think such sites should be on the darkest end of the spectrum. However, the rural sites are mainly lighter, enticing tourists to come and 'enjoy' the location, rather than to reflect and commemorate. The one exception was the rural showcasing of the bushfires in the Macleay region – however, this seemed more targeted at local people rather than attracting tourists. Otherwise, the commemoration and educational aspects were to be found in the urban centres – reflecting the 'civilised' nature of the city, compared to the 'feral' and 'monstrous' nature of the outback. However, it is important to reiterate that tourists travelling to these sites are not 'deviant' – there is a range of motivations for travel to dark sites, and having communities that are actively involved in the storytelling process will ensure control over what narrative is dominant.

Despite the 'lighter' focus of these sites (at times imposed by tour guides, and at others, from the tourists themselves), these locations provide an important vehicle for tourists:

> When visiting dark places tourists can experience a sense of danger and fear, often, mixed with excitement. … Indeed, fear and danger can make people feel alive, and as tourists engage with death and fear from a safe space, they can affectively perceive the grandiosity and magnificence of what happened, which can manifest in an emotion such as excitement, or catharsis. (Martini and Buda, 2020, 685)

As such, the rural provides (often) urban tourists with sites that explore death and destruction in a safe environment. Tourists can venture into the uncivilised outback to witness the horrors of nature (bushfires) or human-induced catastrophes (asbestos) or the dystopian events (murders) against the beautiful landscape of the outback (while experiencing some discomfort from heat, blowflies and of course asbestos fibres), with the knowledge that the experience is transient and that they can return to the safety of the city.

There are of course exceptions to this throughout the chapter; in particular the natural disasters such as flooding and bushfires. While flooding and fires are present in both urban and rural environments, there are often more long-term impacts on rural and regional areas, with rebuilding delayed due to supply issues or agriculture endeavours being devastated.

Further, rural and regional areas are more prone to have their identity recreated based on atrocity or devastation. The ghosts become famous, and the name of the location becomes irrevocably intertwined with the dark past. Wittenoom has been labelled a deadly city; Kapunda has been labelled the most haunted town in Australia, and many locations are now famous for the natural devastation that has occurred. It can be harder for smaller communities to shake these labels, and while some have actively embraced it as a viable economic strategy (for example, Kapunda, although the company itself is urban based, so perhaps the town did not embrace this), others try to reframe and encourage tourists to see what else the community has to offer.

Many tourist sites have moved to an online presence. For some, this has increased recently with new livestreaming events available or online tours and the like. For rural locations, a digital presence offers opportunities to engage tourists who may not be able to travel in person. In these cases, a login requirement provides some financial reimbursement. However, without a personal connection to the site, this may not be the case. Digital platforms also allow dark tourist communities to connect on local, national and international platforms. Tourists are travelling and sharing their experiences more and more online, and communities are growing on sites such as Facebook – and when disaster befalls such communities, there is an increased interest in that place, both physically and digitally. The information available online can also foster false information about a location. For example, in many of the social media sharing of stories of Wittenoom, the dangers of the location are glossed over, and as explored in Chapter 2, false stories are replicated, as in the case of the boab 'prison' trees.

# 8

# Conclusion

In 1869, author Mark Twain published the first American travel book, *Innocents Abroad*, throughout which he documented his 'pleasure trip' across Europe and the Holy Lands with other American travellers. While not all the locations were 'dark', many were, and Twain spent considerable time describing prison cells, morgues, catacombs and sites of assassinations in vivid detail:

> Next we went to visit the Morgue [in Paris], that horrible receptacle for the dead who die mysteriously and leave the manner of their taking off a dismal secret. We stood before a grating and looked through into a room which was hung all about with the clothing of dead men; coarse blouses, water-soaked; the delicate garments of women and children; patrician vestments, hacked and stabbed and stained with red; a hat that was crushed and bloody. On a slanting stone lay a drowned man, naked, swollen, purple; clasping the fragment of a broken bush with a grip which death had so petrified that human strength could not unloose it – mute witness of the last despairing effort to save the life that was doomed beyond all help. A stream of water trickled ceaselessly over the hideous face. We knew that the body and the clothing were there for identification by friends, but still we wondered if any body could love that repulsive object or grieve for its loss. ... I half feared that the mother, or the wife or a brother of the dead man might come while we stood there, but nothing of the kind occurred. Men and women came, and some looked eagerly in, and pressed their faces against the bars; others glanced carelessly at the body, and turned away with a disappointed look – people, I thought, who live upon strong excitements, and who attend the exhibits of the Morgue regularly, just as other people go to see theatrical spectacles every night. When one of these looked in and passed on, I could not help thinking – 'Now this don't afford you any satisfaction – a party with his head shot off is what *you* need.' (Twain, 2003, 92, emphasis in original)

The concluding part of this quote focuses solely on others visiting the same site as a form of 'leisure', with Twain noting that a drowned corpse was not 'intriguing' enough for some of the viewers.

Many of the sites Twain's group visited have now become commodified tourist experiences, attracting international visitors to places such as the Castle d'Ilf, off the coast of Marseille or the Ducal Palace, Venice. Further, many tourists will share similar motivations for travel to such sites – for entertainment, pleasure or titillation. As such, 'pleasure trips' continue in modern society, and indeed 'dark tourism ... is an increasingly pervasive feature in the popular cultural landscape' (Stone, 2009a, 32–33). People frequently travel, locally, nationally and internationally to experience the traumatic past. The motivations of tourists can vary significantly, creating the need for a range of dark tourism providers catering to a range of 'lighter' and 'darker' sites.

For Twain and his fellow sightseers, travelling to Europe and the Holy Lands was restricted to those wealthy enough to embark on the voyage, and it required considerable time and dedication. While most of the sites on his itinerary are now 'urban' centres, the travel to these sites in the 1860s is the equivalent of today's tourists travelling to rural or regional areas (anywhere across the world) to experience dark tourism, or what Twain (2003, 69) labelled 'melancholy history'.

While the narratives presented at rural and regional dark tourism sites are similar to those found in urban areas, rural and regional sites can offer a vastly different experience to the tourist, particularly those locations that have become dystopic among an idyllic setting. The serial killer becomes more 'monstrous' in the outback; the scenic beauty of the Australian bush transforms into hiding places for dangerous (but romanticised) bushrangers or places of threat that can be overcome by selfies where disasters have occurred; old prisons or convict settlements represent the gothic; and colonial sites of violence become folk legends, such as the 'prison tree', as a way to deflect and impose colonial narratives. The remoteness of each location adds *more* experience and *affect* for the tourist by increasing feelings of isolation or heightening natural sounds that may take on a more sinister association at night during ghost tours – tourists are more fully immersed in such locations. There are no (or little) background traffic noises, and no promise of easy retreat to large urban centres. Almost automatically, the rural and regional site adopts an added level of 'darkness' as a result of its geographical location.

This book has presented a series of case studies to highlight the diverse features of rural and regional dark tourism sites. Throughout each case study, a number of common themes have emerged including the possible benefits of dark tourism to remote areas, such as economic and cultural advantages to local communities, the unique constraints or considerations facing tourism providers in geographically isolated areas, the corrupting dystopian influence of dark sites in otherwise idyllic sceneries and the potential for such tourism to be seen as problematic.

# The benefits of rural dark tourism

Dark tourism offers tourists an opportunity to become a spectator to death, violence and atrocity in a 'safe' environment. In all the case studies explored throughout this book, tourists can be educated about Australia's cultural history in an empathetic and (usually) sensitive manner. Further, tourism to such sites offers rural and regional communities the possibility of economic sustainability (or at the least, the chance of increased sustainability). While the motivations of the tourists will influence how a tourist will behave at a dark tourism site, the presence and availability of such sites in rural and regional Australia benefit local communities as well as enabling, reinforcing and/or challenging national cultural heritage.

## Cultural

Cultural criminology encourages researchers to investigate the processes of 'making' and 'attributing' meaning at 'dark' sites – and importantly, to understand the types of narratives that such sites actively promote but also unintentionally convey. The 'dark' sites discussed throughout this book tell us about the social significance of various stages of Australia's history and, importantly, the cultural, social and political shifts in ideology. For example, society shifted the way it viewed convicts, from a 'stain' on society to becoming cultural 'legends' that built Australia. Conversely, bushrangers are changing from being perceived as folklore heroes to 'criminals'. As such, dark tourism can offer many rural and regional locations important cultural benefits – the ability for communities to 'tell their story', as well as sharing and building cultural memory and even national identity. As documented in this book, convicts and bushrangers have a large role to play in the Australian national identity and, as such, offer lots of tourism potential. However, they may have more to offer rural/regional places in terms of building upon (or contrasting) cultural memory than urban areas that have many other 'attractions' to rely upon.

As Chapter 2 highlighted, rural and regional locations have an important role in revealing Australia's history of colonial violence, massacres and the dispossession of land inflicted on Aboriginal people. Shifting political ideologies have seen a reframing of the 'national' story of 'settlement' to one of invasion and dispossession. An important part of this evolving narrative has been encouraging First Nations people to voice their own stories and experiences in public spaces such as museums, art galleries, memorials and tourist sites such as the boab trees. These forms of 'dark tourism' thus provide an important public place to empower those voices that have previously been marginalised and suppressed, as well as provide invaluable cultural information to visitors. Those sites located in rural or regional areas inevitably convey the remoteness of these sites to the visitors, providing a more immersive and empathetic experience for tourists. Urban representations, by comparison, tend to ignore the 'rural' element, and thus an important aspect of Australia's cultural history is overlooked, and visitors (particularly international visitors) may find it difficult to understand the remoteness of many sites of colonial violence.

### Emotional

Martini and Buda (2020) have written about the potential for dark places to have a strong emotional impact on visitors, and this in turn affects how tourists relate and interact with the space. As noted in the introductory chapter, tourists are motivated by a range of desires when visiting 'dark tourism' destinations, such as to further their knowledge, connect with national (or regional) identity, pay tribute or commemorate atrocities, to experience 'thrill, joy, fear, hope, nostalgia and all the embodied experiences and feelings central to these encounters' (Martini and Buda, 2020, 682). In rural and regional areas, there is more opportunity for tourists to be 'affected' by the geographical space and experience the solitude and distance felt by victims of past atrocities. For example, tourists who visit Port Arthur are often struck by the sheer isolation and coldness of the peninsula. These physical sensations are coupled with the knowledge of the hard labour undergone by convicts and the solitary confinement they underwent, adding additional psychological considerations for the tourist. There are minimal urban comforts (with the exception of the tourism centre and gift shop) for tourists to retreat to. Compared to a decommissioned gaol in an urban setting, where the sounds of traffic can still be heard, or the promise of a multitude of coffee shops or accommodation a short walk away, the rural or regional dark tourism site offers tourists more opportunities to be 'affected' by the cultural past.

While many dark tourism sites manipulate experiences to create an 'affect', such as sensations, emotions and bodily actions (Martini and Buda, 2020), the rural or regional atmosphere cannot be easily recreated in urban environments. Thus, the rural and regional landscape offers these destinations a pre-packaged enhancement that may require less manufactured experiences (such as guided tours or jump-scare tactics), although ultimately, many rural dark tourism sites still provide these experiences to tourists, possibly to entice them to travel the required distance. For sites of colonial violence, the rural plays an integral part in immersing visitors in the experience. While urban centres provide tourists with insights into the cultural history of colonial invasion, the contextual geographical information is limited to maps and some images of rural landscapes, making it difficult for tourists to understand how remote some of these sites truly were. Brendan Beirne's infrared photos provide the most compelling evidence of the rurality of these sites of slaughter. Yet, they cannot transport the tourist to the location itself. As such, memorial sites such as Myall Creek provide visitors with a truly immersive and more empathetic experience. That is, individuals can experience the remoteness and the natural setting itself and understand how they may have felt if they were attacked in that location.

## Economic

The growing popularity of dark tourism has led to several United States based studies into the phenomenon. In 2022, Passport Photo Online surveyed 937 Americans about their (dark) tourism experiences. The survey revealed that 82 per cent had visited at least one dark tourism destination in their lifetime and of the remaining 18 per cent of the cohort that had not been to a destination, 63 per cent wanted to (Woolf, 2023). Only 9 per cent of the cohort had a negative view of dark tourism, indicating an overwhelming acceptance of it as a leisure activity.

While the study does not differentiate or focus on the rural or urban location of travel to dark tourism sites, the popularity of dark tourism as a leisure activity indicates the potential economic benefit for communities. Indeed, the estimated global value of the dark tourism market is set to reach US$43.5 billion by 2031 (GlobalData, 2023), and more communities are turning to dark tourism to stimulate the economy. For example, *Visit Ukraine* wanted to provide a 'mid-war experience to travelers' with two main objectives in mind. First, they sought 'to rebuild the fallen tourism industry and accelerate the growth of Ukraine's economy', and second, they sought to 'help the tourist understand the actual situation when they look into the eyes of the residents who had this nerve-wracking experience of war' (Future Market Insights, 2023, para. 6).

Staples (1995, 35) has argued that, as rural and regional economics go through difficult times of change, heritage sites are often turned to as a way of attracting tourists and bringing income into the location. Further, 'tourism is a key element in the new rural economy as rural places are re-imagined, re-packaged and re-presented for a predominantly urban market' (Rofe, 2013, 262). While it is true that Australia has fewer 'dark tourism' sites compared to North America or Europe, the rural and regional landscape offers a rich choice of destinations that can be categorised as 'dark tourist' locations, such as colonial violence memorials, sites of bushranger violence and also decommissioned gaols. Because of the rarity of dark tourism sites in Australia, there is scope and opportunity for rural and regional communities to capitalise on their 'dark' cultural history to create a new tourism market.

Tourism, dark or otherwise, provides revenue for communities, including employing local staff, selling local products such as souvenirs, as well as enriching the wider area through opportunities for accommodation, restaurants and other consumable possibilities. As Gilbert (2006, 187) states, 'entry fees and sales of souvenirs, books and other items contribute directly to the financial income of both the community and its historical enterprises. In turn, historical sites add to the community's ability to attract tourists, encouraging them to spend and to lengthen their stay in the community'.

Not all of the sites explored in this book will bring economic prosperity to a location. For example, it would be difficult to commercialise colonial violence because of its inherent 'darkness'. Yet, for many of the 'lighter' sites, there is the possibility of economic benefits. Shehata et al (2018, 3) have argued that many gaols 'do not stand as examples of conservation or commemoration, but as commercial ventures, especially where they are amenable to public events, open markets, and other social celebrations'. The Old Dubbo Gaol and PAHS have adopted this strategy by encouraging the site as a wedding venue (Old Dubbo Gaol), and as a place where the local community can hold events (Port Arthur). It is also important to note that many of the sites in rural and regional Australia are not financially viable. Port Arthur, for example, requires supplementary support. Despite this, the employment of local people unquestionably provides economic support to the region. Further, employing local people as guides supplements the geographical knowledge imparted to tourists (Crang, 1997, 146).

## Constraints and considerations for establishing rural dark tourism sites

While dark tourism activities may offer rural communities cultural and economic benefits, there can also be antipathy towards such an industry, as seen in the case studies of Snowtown and Belanglo. McKercher (2001, 32) suggests that this antipathy is exacerbated when the tour operator is seen as being an 'outsider' to the community – and this may be even more relevant in close-knit rural/regional communities. Tour operators also need to adopt different strategies for creating and maintaining a successful dark tourist operation within a rural/regional area – for example, they need to ensure that there is sufficient 'material' available to warrant tourists travelling to the area and may consider more targeted 'attractions', such as 'themed' or specialist accommodation. The authenticity of the site and message to visitors is also considered – in particular, whether some sites are providing misconceptions as a way to attract tourists, and whether this may be more likely to occur in rural settings where tourism plays an important part of the economy.

## Community denunciation

One of the key factors affecting a community's acceptance or denunciation of a tragic past relates to the period between the event and the proposed tourism infrastructure. As Shehata et al (2018, 4) note, within dark tourism, it is generally accepted that 'time dilutes associations', and therefore communities are more likely to embrace dark tourism opportunities over time. However, several examples in this book demonstrate that there may be more than 'time passing' to enable rural and regional communities to accept, and embrace, dark tourism activities connected with tragic and atrocious events. Some communities want to distance themselves from the events regardless of the proposed economic benefits. In essence, these places want to revert from the rural dystopia back to the 'ideal' rural community that they were before the traumatic event.

Unfortunately, for rural and regional communities, where the dystopian narrative overtakes all other connotations with the location, it is very difficult to revert to the ideal. Consider the urban case study of Milwaukee, where the Jeffrey Dahmer tours are run. A quick Google search of 'what is Milwaukee best known for' returns results for Tripadvisor (2024), which suggests that Milwaukee is best known for its famous breweries, Major League Brewers and its rich historical and cultural attractions. In contrast, the same search for Snowtown returns numerous pages on the murders and bodies in the barrels.

As such, for those people living in urban areas, while they may not want tourism associated with death or tragedy and may take extreme steps to dissuade such tourism (like demolishing the relevant building and the like), the community has a greater chance of distancing themselves from the event. In rural and regional areas, this often is not possible. For example, the community around Port Arthur tried to reinvent themselves, even renaming the location to Carnarvon, yet convict tourism prevailed. Similarly, in Snowtown, the town community is dedicated to distancing itself from the murders but has so far been unsuccessful in deterring tourists (or changing their online presence). The media have played a large role in this, from the instant media headlines to *Top Gear* and *Virgin Blue* cashing in on the association to create catchy stories and advertisements. Other studies have found that residents are worried about the resale value of their homes where they are located near sites of atrocity (Lennon, 2010), again, this is more likely to impact upon rural and regional communities where there is a reduced competitive market.

The divisive nature of dark tourism is also more likely to impact rural and regional communities than their urban counterparts. While a difference of opinion can lead to disagreements and feuds in all communities, it is more likely to be felt in smaller locations where disagreements become widely known and may lead to residents being 'kicked out of town' like in Snowtown. Throughout the case studies in this book, there have been contrasting and divisive community views as to whether sites should become dark tourism destinations: residents in Uralla stopped talking to one another over the Thunderbolt statue, community members in Goulburn petitioned to have the Ivan Milat tours shut down and the Port Arthur community was divided over both the convict site, and then again, when the massacre occurred.

Residents of sites hosting dark tourism activities are constantly affected by the presence of tourists. As Kim and Butler argue:

> in the context of dark tourism, it is the local community who has to deal with the consequences of representation or (destination) images that are constructed and perceived amongst visitors. Therefore, a local community's participation in dark tourism destination development should be deemed critical. (Kim and Butler, 2015, 80)

An external observer might ask why the residents cannot just distance themselves from any tourism activity, but this is difficult, as tourists often ask residents (especially those working in the service industry) for information on the incident or the tourism. In small towns, this can have ongoing negative impacts on those community members who do not wish to relive the atrocity. In addition, there is also the possibility that relatives of victims (or offenders) remain in the area and will unintentionally (or intentionally) be approached by a tourist for further information. The Port Arthur site has explicitly warned tourists not to ask tour guides or workers about the 1996 massacre exactly for this reason.

## The need for trailscapes?

The potential for rural and regional areas to band together to create trailscapes based on a common theme offers some economic prosperity for a number of communities. The development of the Ned Kelly Touring Route provides tourists with a planned itinerary that promises ongoing tourism infrastructure along the way. That is, tourists embarking on the trail know in advance what each location offers in terms of tourism opportunities, available accommodation and souvenir potential, as well as its 'authenticity' to the Ned Kelly 'story'. As Kim and Butler (2015) noted, in rural and regional areas, tourists must be provided with sufficient 'attractions' and 'creature comforts' to entice visitors out of the city. Those sites that lack such tourism infrastructure encourage more 'intransigent' visitors who are willing to travel vast distances without promises of accommodation or tourism infrastructure.

Bushranger and convict sites have capitalised on trailscapes, encouraging tourists to visit multiple locations that share a common theme. Other locations considered within this book could adopt a similar approach, however, some form of commercialisation would probably be required to make this endeavour economically viable. For example, a website similar to the Ned Kelly Touring Group could be created to promote sites of colonial violence – importantly, such a website could increase the knowledge of colonial violence and provide another platform to give a voice to Aboriginal people. Similarly, where communities want to capitalise on the tragedy of serial killings, a similar approach could be adopted. However, such a trailscape would likely rely heavily upon 'urban' regions with the 'rural' sites connecting 'road-trip' style travel across Australia, or even internationally. There is a clear tourism market, both within Australia and overseas, to visit sites of mass or serial killings in rural areas, as demonstrated by travel to Snowtown and Belanglo Forest, Australia, Plainfield, United States and Soham, United Kingdom.

## Authenticity

Linking back to the notion that many dark tourism sites in rural and regional areas offer cultural links to the past, the issue of authenticity needs to be considered. This is particularly true when such sites are promoting memorialisation over entertainment. For Walby and Piché (2011, 454), 'memorialization in the context of travel and tourism hinges on the idea of authenticity'; that is, travellers want to know that they have seen, touched, or essentially *experienced* history itself through relics, buildings, or even the landscape itself (particularly for bushranger or colonial violence sites). The presentation of authentic exhibits or sites invites audiences to reflect on past practices and to place themselves (safely) in barbaric past events, thus creating the 'affect', or empathetic connection to a site. However, not all sites aim for memorialisation, and some communities that are reliant on economic income from tourism sites may be obliged to provide less authenticity to create a more thrill-seeking environment. Or, in other cases such as Port Arthur, the original buildings may have been destroyed long ago, resulting in the need to recreate spaces as authentically as possible.

With the promise of economic prosperity, some communities may prioritise creating tourist demand as the determinant of their cultural heritage. For example, 'there are tensions between the need for historical enterprises such as living history exhibits and museums to attract tourists and the need for them to be places of memory and identity for community members' (Gilbert, 2006, 186). Port Arthur has tried to maintain this balance by providing free entry to the site to residents – thus ensuring income from 'tourists' while allowing 'locals' to continue to access their heritage. However, this tension also relates to the authenticity of the site and the narrative that is told to tourists. Again, using Port Arthur as a case study, historically the local community did not want tourism and wanted to forget its convict past. The nature of tourism from Hobart meant that the wishes of the local community were overridden and ignored. Similarly, the Ned Kelly Touring Route places so much emphasis on the bushranger, that many other tourist drawcards take a backseat in advertising. These locations no longer house 'bushrangers', yet their history is ever present, particularly in Glenrowan with large statues and recreated scenes, ensuring that the town remains known mainly for its bushranger heritage.

The marketing of local communities based on their history also tends to romanticise the narrative, celebrating what is socially acceptable, and glossing over the more uncomfortable stories. For Gilbert (2006, 193), 'in Australia, heritage tourism, unless managed sensitively, runs the risk of reinforcing a misremembered past in which uncomfortable facts about invasion, dispossession, bigotry and physical and mental cruelty are sanitised'. This has certainly been the case in most sites discussed throughout this book. Port Arthur has changed its narrative several times to keep up to date with current political and social sensibilities but has nonetheless sanitised cruelty at various points. Similarly, the Ned Kelly Touring Route, and sites along the way, have historically presented a nostalgic reminiscence and promoted the Kelly Gang as folk heroes. While this has recently been amended to reintroduce the violence and cruelty of the Kelly Gang, as well as an increased focus on the victims, the overall narrative is still what Gilbert (2006, 193) would term a 'comfortable vision of the past' ensuring that visitors do not experience undue discomfort.

Controlling the narrative is not unique to rural and regional tourism destinations in Australia; it happens in urban environments and across the world. All nations 'continually reinterpret traces of their own history' and use a selected version of history to present 'durable national ideals' (Lowenthal, 1975, 13). The increase in memorials to colonial violence and museum exhibits utilising Aboriginal voices telling Australia's history demonstrate the evolving nature of reinterpreting history, and dark tourism sites change their narratives to reflect progressing understandings of our past. Yet overall, 'Australia continues to celebrate aspects of its colonial and penal history, and Kelly as a romantic symbol of resistance to authority' (Tranter and Donoghue, 2010, 202).

*Environmental issues*

One of the most distinct aspects of rural and regional tourism is the need to consider the impact of tourism on the environment. For the most part, many dark tourism attractions in urban areas are already established locations (penal sites, museums and so on) that are only constrained by the need to expand tourism infrastructure – usually, there is already transport to (or near) the location, and there are fewer constraints on creating further infrastructure as long as it meets local industry codes and heritage constraints (and has sufficient funding). However, in rural and regional areas, there is a number of additional environmental issues that need to be accounted for.

First, as demonstrated through a number of case studies in this book, tourism has led to the destruction of the natural landscape and the travel site itself. The boab trees have been graffitied and the roots trampled, travellers to Wittenoom continually erode the landscape exposing more toxic material and tourists have stolen items from sites to take a memento home – occurring historically at Port Arthur, and more recently at Wittenoom. Media attention focusing on tourism to such sites increases the awareness of the location, in turn encouraging more tourism to the area, thus increasing the risk of environmental damage. In other examples, the natural environment has been reshaped to create tourism sites, infrastructure and/or transport access to such sites, particularly in the more remote locations such as colonial violence memorials or sites of bushranger violence.

With climate change, new environmental issues are threatening rural and regional dark tourist or heritage sites. Port Arthur is in danger from rising sea levels. In the short term, the salt wash is deteriorating the stone and mortar. In the long term, the site could be flooded or land eroded, endangering the site itself (MacDonald, 2023). Port Arthur has already been ravaged and ultimately shaped by two significant bushfires in its past, demonstrating the ongoing fragility of the site. While similar threats are also present in urban areas, there are more services and facilities to prevent damage to urban buildings, and more incentive to prevent environmental damage where surrounding locations will also be affected. In the case of Port Arthur, the site is remote, and the money required to protect the buildings may be deemed unsustainable – this decision was historically made, both when the site was holding convicts and later as a tourism destination. With ongoing changing political narratives and shifts in priorities for funding, this may happen again.

The impact of dark tourism sites on the surrounding environment is something that many rural and regional areas are grappling with worldwide. For example, Sharpley (2009) recounts the ecological concerns surrounding building an international Tsunami Memorial in Khao Lak National Park in Thailand in memory of the catastrophic Indian Ocean tsunami of December 2004. Specifically, concerns were raised about constructing a memorial (that was to include a library, museum, conference rooms, restaurants and space for contemplation and remembrance) within a fragile forest environment and the ongoing impact on the environment with tourist travel. Eventually, the 813 Tsunami Memorial Park was built in Bang Niang, to attract the majority of tourists: while a smaller, more commemorative Baan Nam Khem Tsunami Memorial Park was placed inside Khao Lak featuring plaques for some of the victims.

*Visitor behaviour and motivation*

Several studies have been conducted on visitor behaviour and motivation when visiting dark tourism sites. This growing body of research can be used by tourism providers to determine what type of site to establish, the types of exhibits and narratives to include and the ages to target in marketing. Generally, while all age groups engage in dark travel, younger people (between the ages of 15 and 35) are more likely to travel to dark sites (Future Market Insights, 2023). The key motivations reported by United States travellers to dark sites included that they 'enjoy the educational aspect that comes with dark tourism' (52 per cent); 'wished to pay tribute to people affected by the grief event' (47 per cent); 'wanted to emotionally absorb myself in a place of tragedy' (46 per cent); and finally, 'looking to discover a place with a story rather than just visit a trendy destination' (45 per cent) (Wolfe, 2023, para. 9). It is clear from various research that the motivations for travel differ according to each person, as does the reaction to the site itself.

Many of the types of dark tourism sites discussed in this book were the most appealing for the United States cohort surveyed in 2022: war tourism (56 per cent), disaster tourism (56 per cent), cemetery tourism (53 per cent), ghost tourism (52 per cent), nuclear tourism (50 per cent), Holocaust/genocide tourism (49 per cent) and prison and persecution site tourism (48 per cent) (Woolf, 2023). While a similar study within Australia is likely to show a different order in popularity of dark tourism destinations, this finding provides some evidence for potential tour organisers of the types of activities that are appealing to people, and that marketing could perhaps be used to target an American cohort to attract international tourism.

Of particular concern to rural and regional dark tourism destinations, it was found that where the site is remote, there is a preference to join a tour to 'ensure safety and know more about the destination and related history' (Future Market Insights, 2023, para. 16). As such, while individuals do travel long distances to places such as Wittenoom, tourism to similar places is likely to increase if formal tours are created. The creation of trailscape tours could attract more tourists with promises of learning about a range of different 'dark' sites with the feeling of security offered by an official tour.

# Rural dystopias and tranquil pastoral scenes

Depicting rural areas as dystopian communities can be one way of 'distinguishing' a town and trying to attract more visitors. As Rofe (2013, 262) states 'achieving competitive distinction can be extremely challenging, as the accepted foundation of rural tourism draws significantly upon the discourse of the rural idyll'. The increasing popular culture focus on dark tourism activities, and more broadly on crime, punishment and horrific deaths, has opened up a new avenue for rural towns to market the dystopian landscape of their area. Further, the 'romanticisation' of previous sites of pain and suffering has opened new avenues for tourism. In the example of decommissioned gaols, there has been a romanticisation of the imposing architecture that was once used as a warning to the community of the prisoners it once held inside. Old gaols are no longer threatening, they are now 'romantic', thus making tourists willing to travel and 'pay an admission fee to look over the authentic fabric of former gaols that bear a resemblance to a medieval edifice' (Shehata et al, 2018, 7).

While there is ample evidence of the romanticisation of such sites and the allure of the gothic to tourism, there has been limited analysis of how rural sites present themselves in terms of 'idyll' beauty or 'dystopian' traumascapes, and whether this presentation impacts on the overall 'darkness' of the site. Previous research has highlighted that dark tourist activities can foster negative stereotypes of an area (Kim and Butler, 2015), which can be very disempowering for these communities. Indeed, communities can strive to rebrand themselves as 'idyllic', but outside media references continue to propagate the dystopian stereotypes. Snowtown provides a key example of this – the town has tried many times to re-establish itself as idyllic and distance itself from the 'bodies-in-the-barrels' narrative. Yet, because tourism remains, so too does the allure of profiting from the dystopian element. For example, some local community members want to see dedicated tourism strategies developed around the bank. Similarly, the killings have been used to promote tourism to the entire state of South Australia by large corporations such as airlines to promote their products (and tourism). Each time the killings are promoted as a potential tourism strategy, the town is quick to try to shift the focus back to the 'idyll' nature of the town.

Other communities have been able to embrace the 'dystopian' sites within a broader landscape of beauty and tranquillity – in essence selling a small part of the location as a 'dark tourism' site within a broader context of the 'civilised' Australian 'bush'. Indeed, tourism to an area can lead to communities working towards *improving* the landscape through 'civic beautification programs' (Gilbert, 2006, 190). By way of illustration, the grounds of the PAHS have undergone different stages of 'beautification' throughout its history and the grounds of the site are well known for their tranquil appeal. In addition, as White (2016) notes, the perceptions of 'romantic' landscapes also changed. Where once:

> The horror was in stark contrast to the landscape itself. Though at first seen as gloomy, alien and oppressive, the natural setting soon came to

be regarded as romantically wild, awe-inspiring and picturesque. Tastes were changing. As romanticism seeped into popular consciousness, the idea of wilderness took on new meaning, something to be sought out rather than avoided. The site's neo-Gothic church, badly damaged by fire and covered with ivy, came to be seen as a romantically picturesque ruin. (White, 2016, paras. 14–16)

Recognising the attraction of the natural landscape, the PAHS management has kept the 'dystopian' ruins of the convict settlement and the 1996 massacre and actively worked on surrounding these relics with beautiful sweeping lawns and flowering gardens that encourage tourists to picnic at the site. Souvenir postcards of the site often frame the convict buildings within the broader landscape – incorporating lush green grass, or the harbour, in the picture.

Conversely, the natural beauty of many of these sites diminishes the horror and constricts visitors from confronting their own mortality or questioning past governmental practices. For example, as Tunbridge and Ashworth (1996, 113) have noted, the juxtaposition of the dystopian against the idyll means that 'many rural atrocity sites are tranquil pastoral scenes that do not immediately evoke horror'. Trial Bay Gaol (see Figure 8.1) provides the perfect example of this, where tourists are open about their disbelief that anyone could experience hardship or trauma at such a beautiful location. Here, the rural idyll overcomes the elements of dystopia. Further, the tranquil pastoral scenes can be used to make visitors feel safe and see the sites' atrocities as historical.

## Is dark tourism problematic?

As Woolf (2023) argues, most United States tourists have either travelled to a dark tourism site or want to. Some tourists are also critical of dark tourism. Of the 9 per cent of respondents that disliked dark tourism in the Passport Photo Online survey, 22 per cent believed such tourism exploited human suffering, 18 per cent were concerned with the bias or 'whitewashing' narratives at the sites, 18 per cent believed it was a desecration of human suffering and death, 16 per cent simply did not understand the appeal, 13 per cent viewed it as perverse or inappropriate, and 12 per cent thought it was voyeuristic (Woolf, 2023).

There is some evidence to suggest that 'dark' tourists are also aware of the expected ethical behaviour at sites. For example, 57 per cent of United States respondents disapproved of tourists taking selfies at macabre destinations, and a slightly smaller group (51 per cent) disliked travellers breaking the rules of the dark tourism establishment, highlighting the ethical considerations attached to dark tourism (Woolf, 2023).

Despite the general acceptance of dark tourism by tourists themselves, there is a tendency in the tourism literature to be hyper-critical of tourists

**Figure 8.1:** Aerial photo of Trial Bay Gaol, NSW, Australia

Note: This photograph captures the natural beauty of the location, scenery that leads tourists to disbelief that anyone experienced trauma at this site.

Source: Vyshnya, T. (nd), 'Trial Bay Gaol skyless'

and their behaviour on site, particularly dark tourism sites where tourists may act in ways that seem congruous to the reverence that is normally associated with sites of tragedy, violence and death.

Yet, as Raymen and Smith (2019, 1) argue, 'we live in an age in which "leisure" is viewed as integral to a good life' that enables people to express themselves and their freedom. Further, tourism 'represents a change from normal routines, where accepted behavioral norms do not apply ... and the places visited offer a distinct contrast with the tourist's normal world, and, thus, normal social conventions can be temporarily discarded' (McKercher and du Cros, 2002, 116). Tourism, for most people, provides an escape from normal life, the ability to relax or indulge in the 'other'. Thus, the conversion of sites of tragedy and death into dark dungeons or dark fun factories makes sense.

For some rural and regional areas, it becomes necessary to add a 'dark' entertainment element to a site to attract tourists away from 'urban' centres. The need for entertainment at 'dark' sites is further evidenced by McKercher and du Cros's (2002, 117) observation that 'recreational tourists want or need to experience the novelty of a destination in an enjoyable yet nonthreatening way'. Turning a site into something 'lighter' provides a nonthreatening environment for tourists to 'enjoy' and 'relax' while experiencing dark material. As Robb (2009, 58) argues, 'dark tourism occupies a tense intermediary zone between voyeurism and social justice', and often the way a site is interpreted depends upon the tourists themselves. As demonstrated throughout this book, despite the narrative presented at a particular site, tourists can still choose to look for 'cheap thrills' (Robb, 2009, 58) by 'larking' about at the Port Arthur massacre memorial or taking selfies at Holocaust sites. While many see this behaviour as disrespectful or selfish, others argue that it might be a way to neutralise death or to make it less threatening (Bolan and Simone-Charteris, 2018, 737).

The questioning of travel, and indeed leisure, has a long history. The term 'deviant leisure' reflects the French philosopher Michel Foucault's argument that 'leisure in a society of consumption was a form of idleness and, in turn, a sort of deviation' (Stone, 2013, 85). As such, it is easy to understand why 'dark tourism' is seen as a form of 'deviant leisure' that incorporates 'sensation-seeking behaviour that is immoral, unhealthy or even dangerous' (Stone, 2013, 85). Essentially, such behaviour is (generally) perceived to sit outside the normal bounds of society, where death, violence and atrocity are to be feared and avoided, rather than actively sought out to consume as a leisure activity. Yet, as this book has demonstrated, such 'deviant' tourism has always existed and appears to be growing in popularity within popular culture. The number of 'dark tourism' destinations and experiences is constantly expanding worldwide, and as seen throughout this book, some communities are looking to create 'dark tourism' experiences specifically to attract (international) visitors (for example, Ararat in Victoria, Australia).

## The future of dark tourism

With increased economic growth in the dark tourism industry predicted, and with growing popular culture awareness of the term, the prosperity of dark tourism is assured. As discussed in this book, local governments in places such as Ararat in Victoria are investigating dark tourism strategies to increase tourism and boost the local economy. This is becoming a worldwide trend. For example, the Local Government Information Unit (LGIU) in England created a poll in October 2023 to determine how the community felt about local government involvement in dark tourism. This proactive step into canvassing community support will increase the likelihood of future sites succeeding.

Similarly, the development of tourism to locations affected by the war in Ukraine solidifies emerging trends of a growing interest in disaster tourism. Within Australia, there are other forms of disaster tourism that rural and regional areas could capitalise on more effectively. For example, if the United States tourist survey can be generalised to other nationalities, it is likely that formal (and regulated) travel to sites such as Wittenoom might be profitable, particularly if younger generations are targeted. However, the creation of disaster tourism sites will need to cater to the safety of participants, as well as the protection of the site itself to prevent further natural damage.

Dark tourism sites undoubtedly benefit from a digital presence, particularly those located in rural and regional areas. With an increase in the popularity of dark tourism, sites can capitalise on this by marketing themselves as a dark destination. While an online presence can be beneficial for creating awareness and income through encouraging travel, there are several dangers associated with it. First, the online environment does not care about community feelings or perceptions. Even those locations where communities do not want to be associated with atrocity will likely have a digital presence created by the media or dark tourists themselves via blogs and online groups. Once promoted online or in the media, it is very difficult to move away from such narratives. Second, 'selling' sites of atrocity online risks desensitising the material in their pursuit of attracting tourists. The Belanglo State Forest tours is a prime example of this, where the marketing of these tours focused on 'fright' tourism and sensational claims of the potential to find more bodies, causing outrage among the community and the relatives of Milat's victims regarding the insensitive approach to the tours. As such, tourism sites need to present their location in an 'authentic', empathetic manner that encourages tourists to travel into rural and regional environments.

Dark tourism sites could also adopt the Port Arthur strategy of outlining what is expected of visitors from the time of their arrival. With increased media attention to dark tourism, it is more likely that families and friends of victims, and the community more broadly, will be retraumatised with ongoing media fascination and tourists that are after 'extra' information (or the smell of dead bodies as in Snowtown). As such, working with the local community to determine boundaries around acceptable behaviour will aid in protecting community members as well as prevent unauthorised tourism ventures.

However, this assumption relies upon the premise that communities will ultimately engage in tourism strategies. For those communities that do not want dark tourism, the media can help to disseminate this message, but at the same time, it ultimately promotes something that is seen as enticing and 'off limits', or as argued in the case of Wittenoom, sending messages that tourism is not wanted is like 'having a bag of lollies in the cupboard and saying, "Kids, don't eat these lollies"' (Bodie Norman cited in Birch and Gorman, 2021, para. 27).

To this, there is no easy solution except to appeal to tourists to respect a community's wishes and, at the least, act responsibly and empathetically while at the site. As dark tourism becomes more normalised (rather than portrayed as deviant), it is likely that this will naturally follow, as demonstrated by the United States online survey that indicated that over half of respondents disliked inappropriate behaviour at dark tourism sites. As such, there is a growing awareness of the ethical issues surrounding dark tourism and the need to be sensitive when engaging in tourism practices.

## The significance of rural dark tourism

Dark tourism is a way to remember (and at times to celebrate) darker aspects of a community's or nation's heritage and culture. As such, tour operators have an important role (and ethical responsibility) in defining how rural/regional areas are constructed and portrayed at sites. As Sharpley and Stone (2009, 250) note, each 'dark tourism' site provides a 'multitude of meanings and purposes with respect to both their production and consumption' that will, inevitably, be 'consumed' in different ways by tourists, depending on their own background and expectations. As such, no site remains static along Stone's 'spectrum of darkness', or indeed, is the same for all tourists.

The introduction to this book discussed the limitations of dark tourism as a theoretical framework and highlighted some of the gaps in knowledge, such as 'what is so "dark" about dark tourism?' (Bowman and Pezzullo, 2010, 188), and what makes a site 'dark' compared to 'heritage'? Each of the case studies explored throughout this book has described the 'darkness' inherent with each location, yet many places also remain 'heritage' sites. A site can be both a heritage site and a dark tourism site. The difference perhaps lies with the intention of the tourist in visiting the site – are they going for memorialisation purposes or to engage in thrill-seeking and titillating behaviour? Some sites are openly embracing the 'dark' label, simply to attract tourism, while others are trying to move away from any association with such tourism.

It is clear that the 'lighter' dark sites do tend to attract more tourism and, as such, have higher levels of tourism infrastructure and souvenir opportunities. For example, most of the tourism associated with convicts, bushrangers and penal sites was aimed at families, and the kitsch souvenirs offered reflect the 'lightness' of the site and the focus on entertainment. These sites had a more detailed online presence, with targeted marketing strategies that focused on 'fascinating stories' of violence, crime and death. In comparison, the 'darker' sites, such as colonial violence sites, natural disasters and serial killers were more 'memorial' in nature, refraining from (official) souvenirs and only promoting events such as commemorations. Despite this, such sites still attract 'dark' tourists, many travelling vast distances to experience the site. In the example of Wittenoom, many travellers appeared to be thrill seekers – engaging in risk-taking behaviour in pursuit of the perfect selfies, while others seemed to be attracted to (inaccurate) myths such as with the boab prison trees.

The examples in this book indicate that the more 'rural' the location, the less tourism infrastructure there was. However, this is also linked to the 'darkness' of the site – many of the most remote locations shared in this book are sites of colonial violence or disasters, perhaps limiting the opportunity to commercialise these cultural places. Despite the lack of infrastructure, some places, such as Snowtown, continue to attract visitors keen to view a site of death and atrocity.

While dark tourism has been criticised as the 'dirty little secret of the tourism industry' (Marcel, 2004, 1, cited in Stone, 2009b), the 'Disneyfication' of tragedy may be beneficial for *some* sensitive heritage sites. For example, providing effective and engaging thematic storytelling that guides visitors through a meaningful experience can provide both education and entertainment (Heidelberg, 2015, 76) in an empathetic and affective manner. As such, dark tourism has the potential to promote awareness and understanding of historical (or more current) issues and serve to stimulate the economy of local areas in both Australia and internationally. However, as discussed throughout this book, dark tourism is not always welcome, particularly when it has a high entertainment focus. Some communities 'see it as unwelcome sensationalism and having fun at the expense of death', that ultimately exploits victims and forever connects their 'home territory' with tragedy and dystopia (Heidelberg, 2015, 78). Unfortunately, the negative effects of dark tourism are more likely to impact rural and regional areas where there may be fewer sites to attract tourists to visit, or to distract the negative media attention that tends to focus on the 'dark' elements. Rural locations are more likely to be negatively stigmatised by tragedy, violence and death, and ultimately remain as rural dystopias despite community efforts to avoid such connections.

# References

ABC News (2015) 'Ivan Milat: victims' group hits out at terror tour of Belanglo State Forest where killer brutally murdered, buried seven people', *ABC News* [online] 14 July, Available from: https://www.abc.net.au/news/2015-07-14/belanglo-tour-of-ivan-milat-killing-ground-slammed/6617660 [Accessed 3 May 2022].

Ace (1888) 'The remains of a convict settlement', *Argus*, 2 June 1888, page 4.

Addeo, F., Punziano, G. and Padricelli, G.M. (2021a) 'Using digital methods to shed light on "border phenomena": a digital ethnography of dark tourism practices in time of COVID-19', *Italian Sociological Review*, 11(4S): 269–291.

Addeo, F., Punziano, G. and Padricelli, G.M. (2021b) 'Prohibitions, pleasures, and disasters: entering the online "red zone" as an experience of digital dark tourism in time of COVID-19', *Culture e Studi del Sociale*, 6(1), Special: 211–218.

Advertiser (2000) 'Town's bid to move on', *Advertiser* [Adelaide, South Australia, Australia], 14 August: 007.

Advertiser (2004) '"Barrel" ads withdrawn', *Advertiser* [Adelaide, South Australia, Australia], 26 March: 005.

Age (1924) 'The Hobart Carnival: Victorians visit Port Arthur', *Age*, 12 August: 11.

Alpine Helicopter Charter (2021) *Discover Our Scenic Tours*, Available from: https://www.alpineheli.com.au/packages [Accessed 31 August 2022].

Ararat Asylums (nda) 'About J Ward', *J Ward Lunatic Asylum*, Available from: https://www.araratasylums.com/about-j-ward-aradale-1 [Accessed 8 December 2021].

Ararat Asylums (ndb) 'J Ward Historical Day Tour', *J Ward Lunatic Asylum*, Available from: https://www.araratasylums.com/j-ward-day-tour [Accessed 8 December 2021].

Ararat Asylums (ndc) 'J Ward Lunatic Asylum Ghost Tour', *J Ward Lunatic Asylum*, Available from: https://www.araratasylums.com/j-ward-ghost-tour [Accessed 8 December 2021].

Ararat Asylums (ndd) 'J Ward Lunatic Asylum 3 hour paranormal investigation', *J Ward Lunatic Asylum*, Available from: https://www.araratasylums.com/j-ward-overnight-investigation [Accessed 8 December 2021].

Archibald-Binge, E. (2021) 'Unsettled, an Indigenous-led exhibition at the Australian Museum, unearths Australia's "uncomfortable history"', *ABC News*, [online] 18 May, Available from: https://www.abc.net.au/news/2021-05-18/indigenous-exhibition-on-australias-uncomfortable-history/100143944 [Accessed 18 May 2021].

Argo (1912) 'Old time reminiscences: Port Arthur seventy-six years ago', *Critic*, 3 February: 4.

Argus (1890) 'A holiday cruise in the Pateena: relics of the convict system, II Port Arthur', *Argus*, 11 April: 6.

Argus (1890) 'A holiday cruise in the Pateena: the relics of the convict system, III Leaving Port Arthur', *Argus*, 15 April: 9.

Asbestos Diseases Society of Australia Inc (nd) *The Wittenoom Tragedy*, Available from: https://asbestosdiseases.org.au/the-wittenoom-tragedy/ [Accessed 28 September 2022].

Atlas Obscura (2021) 'Backpackers memorial', *Atlas Obscura*, Added by VeloObscura; edited by akohler726, IJVin, 15 October, Available from: https://www.atlasobscura.com/places/backpackers-memorial [Accessed 3 May 2022].

Aussie Towns (2021) *Wittenoom, WA*, Available from: https://www.aussietowns.com.au/town/wittenoom-wa [Accessed 27 September 2022].

Australasian Sketcher (1881) 'School picnic on the Glenrowan battle-field', *Australasian Sketcher*, 24 September: 305.

Australian Bureau of Statistics (1962) *Census of the Commonwealth of Australia, 30th June, 1961*, Canberra: Australian Bureau of Statistics, Available from: https://www.ausstats.abs.gov.au/ausstats/free.nsf/0/E9031BF823C21441CA2578790019D3B9/$File/1961%20Census%20-%20Volume%20V%20-%20Part%20V%20WESTERN%20AUSTRALIA%20Population%20and%20Dwellings%20in%20Localities.pdf [Accessed 27 September 2022].

Australian Government Sydney Harbour Federation Trust (2021) 'Captain Thunderbolt: legendary bushranger', *Harbour Trust*, Available from: https://www.harbourtrust.gov.au/en/our-story/harbour-history/digitales/captain-thunderbolt/ [Accessed 27 September 2022].

Australian Museum (2021a) 'Unsettled introduction', *Australian Museum*, Available from: https://australian.museum/learn/first-nations/unsettled/unsettled-introduction/ [Accessed 1 March 2022].

Australian Museum (2021b) 'Fighting wars', *Australian Museum*, Available from: https://australian.museum/learn/first-nations/unsettled/fighting-wars/ [Accessed 1 March 2022].

Australian Museum (2021c) 'The approach to the Warrego Country map, c. 1845', *Australian Museum*, Available from: https://australian.museum/learn/first-nations/unsettled/fighting-wars/warrego-country-map/ [Accessed 1 March 2022].

Australian Museum (2021d) 'The Sydney wars', *Australian Museum*, Available from: https://australian.museum/learn/first-nations/unsettled/fighting-wars/sydney-wars/ [Accessed 1 March 2022].

Australian Museum (2021e) 'The Appin Massacre', *Australian Museum*, Available from: https://australian.museum/learn/first-nations/unsettled/fighting-wars/appin-massacre/ [Accessed 1 March 2022].

Australian Museum (2021f) 'Remembering massacres', *Australian Museum*, Available from: https://australian.museum/learn/first-nations/unsettled/remembering-massacres/ [Accessed 1 March 2022].

Australian Museum (2021g) 'Map of colonial frontier massacres in Australia 1788–1930', *Australian Museum*, Available from: https://australian.museum/learn/first-nations/unsettled/remembering-massacres/map-of-colonial-frontier-massacres/ [Accessed 1 March 2022].

Australian Museum (2021h) 'Missions, reserves and stations', *Australian Museum*, Available from: https://australian.museum/learn/first-nations/unsettled/surviving-genocide/missions-reserves-stations/ [Accessed 1 March 2022].

Australian Museum (2021i) 'Stolen generations', *Australian Museum*, Available from: https://australian.museum/learn/first-nations/unsettled/surviving-genocide/stolen-generations/ [Accessed 1 March 2022].

Australian Museum (2022a) 'The Australian Museum's Unsettled exhibition', *Australian Museum*, Available from: https://australian.museum/learn/first-nations/unsettled/ [Accessed 1 March 2022].

Australian Museum (2022b) 'Dark Days: a photo essay by Brendan Beirne', *Australian Museum*, Available from: https://australian.museum/learn/first-nations/dark-days-brendan-beirne/ [Accessed 1 March 2022].

Australian Paranormal Phenomenon Investigators (APPI) (nd) 'Berrima Court House ghost hunts', *APPI Ghost Hunts & Tours*, Available from: https://www.appighosthunts.com/berrima-court-house-ghost-hunts.html [Accessed 13 December 2021].

Australia's North West (2022) 'Boab prison tree', *Australia's North West*, Available from: https://www.australiasnorthwest.com/explore/kimberley/derby/boab-prison-tree#no-back [Accessed 21 February 2022].

Australian Town and Country (1880a) 'Further particulars', *Australian Town and Country*, 3 July: 8.

Australian Town and Country (1880b) 'Excitement at Glenrowan', *Australian Town and Country*, 3 July: 8.

Barnard, E. (2010) *Exiled: The Port Arthur Convict Photographs*, Canberra: National Library of Australia.

Barnes, J. and McIntyre, J. (2017) '"A funny place" for a prison: coastal beauty, tourism, and interpreting the complex dualities of Trial Bay Gaol, Australia', in J.Z. Wilson, S. Hodgkinson, J. Piché and K. Walby (eds) *The Palgrave Handbook of Prison Tourism*, London: Palgrave, pp 55–83.

Beatties Studio (1990) *The Convict Days of Port Arthur* (4th edn), Hobart: Beattie's Studio.

Beeton, S. (2004) 'Rural Tourism in Australia – has the gaze altered? Tracking rural images through film and tourism promotion', *International Journal of Tourism Research*, 6: 125–135.

Begg, P. (2005) *Jack the Ripper: The Definitive History*, London: Pearson Educational Group.

Bennett, C. (2015) 'The blue ghosts of Wittenoom', *WA Today*, Available from: https://www.watoday.com.au/interactive/2015/blueGhosts/ [Accessed 28 September 2022].

Berrima Courthouse Trust (2021a) 'The history', *Berrima Courthouse*, Available from: https://berrimacourthouse.org.au/about/ [Accessed 13 December 2021].

Berrima Courthouse Trust (2021b) *Berrima Courthouse*, Available from: https://berrimacourthouse.org.au [Accessed 13 December 2021].

Berrima Courthouse Trust (2021c) 'The courthouse today', *Berrima Courthouse*, Available from: https://berrimacourthouse.org.au/about/cou rthouse-today/ [Accessed 13 December 2021].

Berrima Courthouse Trust (2021d) 'Tours', *Berrima Courthouse*, Available from: https://berrimacourthouse.org.au/tours/ [Accessed 13 December 2021].

Berrima Courthouse Trust (2021e) 'Visitors', *Berrima Courthouse*, Available from: https://berrimacourthouse.org.au/visitors/ [Accessed 13 December 2021].

Berrima District Historical and Family History Society (2012a) 'Tourism efforts led to visitors centre', *Southern Highland News*, [online] 15 October, Available from: https://www.southernhighlandnews.com.au/story/395 785/tourism-efforts-led-to-visitors-centre/ [Accessed 13 December 2021].

Berrima District Historical and Family History Society (2012b) 'Much more than just a courthouse', *Southern Highland News*, [online] 5 February, Available from: https://www.southernhighlandnews.com.au/story/1078 439/much-more-than-just-a-court-house/ [Accessed 13 December 2021].

Berrima Village (2013) 'Historic Berrima', *Historic Berrima Village*, Available from: http://berrimavillage.com.au/historic-berrima/ [Accessed 13 December 2021].

Birch, L. and Gorman, V. (2021) 'Why are people still travelling to asbestos-riddled Wittenoom?', *ABC News*, [online] 12 August, Available from: https://www.abc.net.au/news/2021-08-12/asbestos-riddled-wittenoom-draws-visitors-despite-health-risks/100369764 [Accessed 28 September 2022].

Birch, T. (1999) 'Come see the giant koala', *Meanjin*, 58(3): 60–72.

Black, J.M. (1890) 'Letters to the editor: Port Arthur', *Mercury*, 17 January: 4.

Boggo Road Gaol Historical Society (nd) 'Prison slang', *Inside Boggo Road: The History of Queensland Prisons*, Available from: https://www.boggoroadgaol.com.au/2015/10/prison-slang.html [Accessed 30 November 2022].

Bogle, M. (2008) *Convicts: Transportation and Australia*. Sydney: Historic Houses Trust of New South Wales.

Bolan, P. and Simone-Charteris, M. (2018) 'Shining a digital light on the dark: harnessing online media to improve the dark tourism experience', in P.R. Stone, R. Hartmann, T. Seaton, R. Sharpley and L. White (eds) *The Palgrave Handbook of Dark Tourism Studies*, London: Palgrave Macmillan, pp 727–746.

Bowman, M.S. and Pezzullo, P.C. (2010) 'What's so "dark" about "dark tourism"?: death, tours, and performance', *Tourist Studies*, 9(3): 187–202.

Bradtke, B. (2022) 'The boab tree – our Kimberley icon', *Outback Australia Travel Secrets*, Available from: https://www.outback-australia-travel-secrets.com/boab_tree.html [Accessed 7 December 2022].

Brand, I. (2003) *The "Separate" or "Model" Prison, Port Arthur*, Launceston: Regal Publications.

'Bridle Track' [pseudonym] (1950) 'Westraliana fauna, facts, flora. We take a closer look at Ravensthorpe in the Boom Days: the smelters', *Western Mail*, 1 June: 14.

Bristow, R. and Newman, M. (2004) 'Myth vs. fact: an exploration of fright tourism', *Proceedings of the 2004 Northeastern Recreation Research Symposium*, Available from: https://www.fs.usda.gov/ne/newtown_square/publications/technical_reports/pdfs/2005/326papers/bristow326.pdf [Accessed 19 November 2021].

Brook, H. (2009) 'Dark tourism', *Law Text Culture*, 13: 259–272, Available from: http://ro.uow.edu.au/ltc/vol13/iss1/12 [Accessed 19 November 2021].

Brooks, A. (1964) *Tree Wonders of Australia*, Melbourne: Heinemann.

Brookes, S. (2022) 'Last remaining Wittenoom properties to be demolished in bid to deter danger-seeking tourists', *WAtoday*, [online] 24 March, Available from: https://www.watoday.com.au/national/western-australia/last-remaining-wittenoom-properties-to-be-demolished-in-bid-to-deter-danger-seeking-tourists-20220324-p5a7ju.html [Accessed 28 September 2022].

Brown, J. (2013) 'Dark tourism shops: selling "dark" and "difficult" products', *International Journal of Culture, Tourism and Hospitality Research*, 7(3): 272–280.

Brown, M. (2009) *The Culture of Punishment: Prison, Society and Spectacle*, New York: New York University Press.

Bulbeck, C. (1991) 'Aborigines, memorials and the history of the frontier', *Australian Historical Studies*, 24(96): 168–178.

Bunn, A. (2007) 'Plaque theft angers police', *The Border Mail*, [online] 3 September, Available from: https://www.bordermail.com.au/story/36291/plaque-theft-angers-police/ [Accessed 31 August 2022].

Byrnes, J. (2022) 'Museum of Australian policing: educating the community', *Australian Police Journal*, 4 June, Available from: https://apjl.com.au/museum-of-australian-policing/ [Accessed 15 May 2023].

Canberra Times (1976) 'Thief makes off with Kelly's colt', *Canberra Times*, 20 September: 3.

Carroll, J. (1992) *Intruders in the Bush: The Australian Quest for Identity* (2nd edn), Melbourne: Oxford University Press.

Carroll, N. (2020) ' "Australia's Chernobyl": Instagram users warned over outback "ghost town"', *Yahoo!News*, [online] 11 December, Available from: https://au.news.yahoo.com/instagram-users-warned-over-outback-ghost-town-003359341.html [Accessed 28 September 2022].

Casella, E.C. and Fennelly, K. (2016) 'Ghosts of sorrow, sin and crime: dark tourism and convict heritage in Van Diemen's Land, Australia', *International Journal of Historical Archaeology*, 20(3): 506–520.

Casella, E.C. and Fredericksen, C. (2004) 'Legacy of the "fatal shore": the heritage and archaeology of confinement in post-colonial Australia', *Journal of Social Archaeology*, 4(1): 99–125.

Cave, J. and Buds, D. (2018) 'Souvenirs in dark tourism: emotions and symbols', in P.R. Stone, R. Hartmann, T. Seaton, R. Sharpley and L. White (eds) *The Palgrave Handbook of Dark Tourism Studies*, London, Palgrave Macmillan, pp 707–726.

Change.org (2015) 'Stop the Belanglo State Forest Ivan Milat terror tour and make Goulburn Ghost Tours issue an apology to the families of the victims', *change.org*, Available from: https://www.change.org/p/nsw-premier-mike-baird-goulburn-ghost-tours-stop-the-belanglo-state-forest-ivan-milat-terror-tour-and-make-goulburn-ghost-tours-issue-an-apology-to-the-families-of-the-victims [Accessed 3 May 2022].

Chen, L. (2018) *Glenrowan Heritage Precinct: Conservation and Landscape Management Plan*, Victoria: Rural City of Wangaratta.

Chernobyl Tour (2021a) *Chernobyl Zone Guest Code*, Available from: https://chernobyl-tour.com/chernobyl_zone_guest_code.html [Accessed 12 October 2022].

Chernobyl Tour (2021b) *Air of Chernobyl and Pripyat – Canned Souvenirs from Chornobyl Zone*, Available from: https://chernobyl-tour.com/air_of_chernobyl_pripyat_canned_en.html [Accessed 12 October 2022].

Church of England (1890) 'A trip to Port Arthur (Tasmania)', *Church of England Messenger for Victoria and Ecclesiastical Gazette for the Diocese of Melbourne*, 7 March: 262.

Clarence and Richmond Examiner (1906) 'Tourists in the Macleay district', *Clarence and Richmond Examiner*, 10 November: 4.

Cleary, J. (1955) 'A scream of terror and then murder', *Argus*, 3 December: 10.

Cochrane, P. (2022) 'Ways of seeing: photographs as historical evidence', *The National Centre for History Education*, Australian Government, Available from: https://hyperhistory.org/index.php?option=displaypage&Itemid=574&op=page [Accessed 28 September 2023].

Cock, A. (1999) 'The missing people murders; ghouls already passing through', *Advertiser* [Adelaide, South Australia, Australia], 24 May: 006.

Coia, A. (2012) 'The Ned Kelly trail', *Weekend Notes*, [online] 29 October, Available from: https://www.weekendnotes.com/the-ned-kelly-trail-melbourne/ [Accessed 31 August 2022].

Coleman, M. (2000) 'Snowtown struggles to change image', *ABC News*.

Collins, N. (2021a) 'Overview of H.M. Prison Beechworth', *HM Gaol Beechworth*, Available from: https://hmgaolbeechworth.com/Home/Timeline [Accessed 10 December 2021].

Collins, N. (2021b) 'Deaths in custody', *HM Gaol Beechworth*, Available from: https://hmgaolbeechworth.com/Home/DeathsInCustody [Accessed 10 December 2021].

Context Pty Ltd, Urban Initiatives Pty Ltd and Doyle, H. (2002) *Landscape Management Plan, Context: Community, Heritage and Environment Solutions*, Available from: https://portarthur.org.au/wp-content/uploads/2017/12/PAHSMA-Landscape-Management-Plan.pdf [Accessed 10 May 2019].

Corbin, A., Hunt, W.J., Vavano, C. and Harris, M.J. (2010) 'The Marshall/Firehole Hotel: archaeology in a thermal river environment', in A. Corbin and M.A. Russell (eds) *Historical Archaeology of Tourism in Yellowstone National Park*. New York: Springer, pp 159–225.

Cornish, R. (2016) 'Glenrowan and the Warby ranges: Ned Kelly country', *Sydney Morning Herald*, [online] 19 May, Available from: https://www.smh.com.au/entertainment/glenrowan-and-the-warby-ranges-ned-kelly-country-20160518-goy0mf.html [Accessed 30 August 2022].

Crang, P. (1997) 'Performing the tourist product', in C. Rojeck and J. Urry (eds) *Touring Cultures: Transformations of Travel and Theory*. London and New York: Routledge, pp 137–154.

Crittenden, V. (1976) 'Ward, Frederick (Fred) (1835–1870)', *Australian Dictionary of Biography*, Volume 6, Melbourne: Melbourne University Press, Available from: https://adb.anu.edu.au/biography/ward-frederick-fred-4801/text8001, [Accessed 5 September 2022].

Dale, C. and Robinson, N. (2011) 'Dark tourism', in P. Robinson, S. Heitmann and P.U.C. Dieke (eds) *Research Themes for Tourism*, Wallingford: CABI, pp 205–217.

Dalton, D. (2015) *Dark Tourism and Crime*, London: Routledge.

Daniels, K. (1983) 'Cults of nature, cults of history', *Island Magazine*, September: 3–8.

Davies, P. (2004) *Trial Bay Goal: Conservation Management and Cultural Tourism Plan*, Volume 1. Sydney, NSW, Available from: https://webarchive.nla.gov.au/awa/20090401020000/http://pandora.nla.gov.au/pan/97622/20090 401-0940/www.environment.nsw.gov.au/resources/parks/CMPArakoon SCATrialBayGaol.pdf [Accessed 12 May 2023].

De Poloni, G. (2018) 'Tourists warned over visiting asbestos-riddled Wittenoom, Australia's most dangerous ghost town', *ABC News*, [online] 12 July, Available from: https://www.abc.net.au/news/2018-07-12/austral ias-deadliest-ghost-town-wittenoom-a-tourist-hotspot/9987328 [Accessed 28 September 2022].

Debelle, P. (2019) 'From the archives, 1999: up to six bodies found in barrels in Snowtown', *The Sydney Morning Herald*, [online] 17 May, Available from: https://www.smh.com.au/national/from-the-archives-1999-up-to-six-bodies-found-in-barrels-in-snowtown-20190517-p51ojx.html [Accessed 26 May 2022].

Department of Agriculture, Water and the Environment (2021) *Australian Convict Sites*, Available from: https://www.awe.gov.au/parks-heritage/herit age/places/world/convict-sites [Accessed 18 October 2021].

Design 5 – Architects Pty Ltd (2003) *The Separate (Model) Prison Port Arthur: Conservation project report*, Available from: https://portarthur.org.au/wp-content/uploads/2017/12/Separate-Prison-Conservation-Proj ect-Report.pdf [Accessed 10 May 2019].

Dubbo Regional Council (2021a) 'Discover', *Old Dubbo Gaol*, Available from: https://www.olddubbogaol.com.au/discover [Accessed 14 December 2021].

Dubbo Regional Council (2021b) *Old Dubbo Gaol*, Available from: https://www.olddubbogaol.com.au [Accessed 14 December 2021].

Dubbo Regional Council (2021c) 'Visit', *Old Dubbo Gaol*, Available from: https://www.olddubbogaol.com.au/visit [Accessed 14 December 2021].

Dubbo Regional Council (2021d) 'Learn', *Old Dubbo Gaol*, Available from: https://www.olddubbogaol.com.au/learn [Accessed 14 December 2021].

Dunn, J. (2008) 'The town that won't die', *R.M Williams Outback Magazine*, Issue 58, Available from: https://www.outbackmag.com.au/the-town-that-wont-die/ [Accessed 17 May 2023].

Edge Insights (nd) *Ararat Visitor Economy Strategy 2018–2021*, Available from: https://www.ararat.vic.gov.au/sites/default/files/document/Ara rat%20Visitor%20Economic%20Strategy%202018-2021%20web.pdf [Accessed 8 February 2022].

Edmonds, L. (2019) 'The Gaol on the hill: the prelude to and construction of Bendigo's sandstone gaol', *Before/Now*, 1(1): 47–58.

Evening News (1939) 'Convict horrors only lies of souvenir sellers: Port Arthur historians claim', *Evening News*, 19 May: 4.

Fagence, M. (2016) 'Tourism-historic significance of small townships in a dispersed spatial pattern: speculations using an Australian example', *Tourism Spectrum*, 2(1): 11–20.

Fagence, M. (2017) 'A heritage 'trailscape': tracking the exploits of historical figures – an Australian case study', *Journal of Heritage Tourism*, 12(5): 452–462.

Ferguson, M., Piché, J. and Walby, K. (2019) 'Representations of detention and other pains of law enforcement in police museums in Ontario, Canada', *Policing and Society*, 29(3): 318–332.

Ferrell, J., Hayward, K. and Young, J. (2008) *Cultural Criminology: An Invitation*, London: SAGE Publications.

Fewster, S. and Innes, S. (2011) 'Killing Adelaide's murder myth', *The Advertiser*, [online] 16 February, Available from: http://www.adelaidenow.com.au/news/killing-adelaides-murder-myth/news-story/fb572436b4320acc7d36e9988c7fa2e1?sv=ae8effb9bfe9e3a477f1836cdcfab47 [Accessed 26 May 2022].

Foote, K.E. (2003) *Shadowed Ground: America's Landscape of Violence and Tragedy*, Austin: University of Texas Press.

Foster, M. (2020) 'Texture, light and sound: a sensory history of early Sydney. Hyde Park Barracks Museum, Sydney', *Australian Historical Studies*, 51(3): 344–347.

Frew, E.A. (2012) 'Interpretation of a sensitive heritage site: the Port Arthur Memorial Garden, Tasmania', *International Journal of Heritage Studies*, 18(1): 33–48.

Friends of J Ward (2017a) 'J Ward wins Victoria museum award', *J Ward*, Available from: https://www.jward.org.au/stories/j-ward-wins-victorian-museum-award/ [Accessed 8 December 2021].

Friends of J Ward (2017b) *J Ward*, Available from: https://www.jward.org.au [Accessed 8 December 2021].

Frow, J. (1999) 'In the penal colony', *Australian Humanities Review*, 13, Available from: http://australianhumanitiesreview.org/1999/04/01/in-the-penal-colony/ [Accessed 1 August 2019].

Future Market Insights (2023) *Dark Tourism Market*, Available from: https://www.futuremarketinsights.com/reports/dark-tourism-sector-overview [Accessed 12 February 2024].

Gibson, D.C. (2006) 'The relationship between serial murder and the American tourism industry', *Journal of Travel & Tourism Marketing*, 20(1): 45–60.

Gilbert, S. (2006) 'Finding a balance: heritage and tourism in Australian rural communities', *Rural Society*, 16(2): 186–198.

Ghost Crime Tours (2019a) *Kapunda Ghost Crime Tour*, Available from: https://www.ghost-crime-tours.com.au/kapunda-ghost-crime-tour [Accessed 22 November 2022].

Ghost Crime Tours (2019b) North *Kapunda Hotel: Paranormal Lockin*, Available from: https://www.ghost-crime-tours.com.au/north-kapunda-hotel-paranormal-lockin [Accessed 22 November 2022].

Glenrowan Tourist Centre (2022) 'The show', Available from: https://www.glenrowantouristcentre.com.au/the-show/ [Accessed 30 August 2022].

GlobalData (2023) 'Shedding light on dark tourism', *Verdict*, [online] 29 August, Available from: https://www.verdict.co.uk/growth-of-dark-tourism/?cf-view&cf-closed [Accessed 12 February 2024].

Goc, N. (2002) 'From convict prison to the gothic ruins of tourist attraction', *Historic Environment*, 18(3): 22–26.

Golańska, D. (2015) 'Affective spaces, sensuous engagements: in quest of a synaesthetic approach to "dark memorials"', *International Journal of Heritage Studies*, 21(8): 773–790.

Goodfellow, H. (2003) 'Snowtown life goes on despite rubbernecks', *The Mercury* (Tasmania), 10 September: 12.

Gordon, B. (1986) 'The souvenir: messenger of the extraordinary', *Journal of Popular Culture*, 20(3): 135–146.

Gordon, R.M. (2001) *Alias Jack the Ripper*, Jefferson, NC: McFarland Publishing Company.

Goulburn Ghost Tours (2015) 'GGT extreme terror tour: will you survive?', *Goulburn Ghost Tours*, website no longer available [Accessed 14 July 2015].

Grant, E. and Harman, K. (2017a) 'Dark tourism, Aboriginal imprisonment and the 'prison tree' that wasn't', *The Conversation*, [online] 28 March, Available from: https://theconversation.com/dark-tourism-aboriginal-imprisonment-and-the-prison-tree-that-wasnt-75203 [Accessed 2 December 2022].

Grant, E. and Harman, K. (2017b) 'Inventing a colonial dark tourism site: the Derby boab "prison tree"', in J.Z. Wilson, S. Hodgkinson, J. Piché and K. Walby (eds) *The Palgrave Handbook of Prison Tourism*, London: Palgrave, pp 735–759.

Grapho (1925) '"Way down south": a trip to historical Port Arthur', *Advocate*, 16 May: 12.

G.T.H. (1910) 'Port Arthur', *Critic*, 5 February: 2.

Halkett, J. (2017) 'Baobab trees – upside-down giants', *Talking Trees*, [online] 3 June, Available from: http://www.talkingtrees.com.au/baobab-trees-upside-down-giants/ [Accessed 2 December 2022].

Harris, B. and Wise, J. (2018) 'Capturing crime in the Antipodes: Colonist cultural representation of Indigeneity', in K. Carrington, R. Hogg, M. Sozzo and J. Scott (eds) *Palgrave Handbook of Criminology and the Global South*, London: Palgrave Macmillan, pp 391–413.

Harris, J., Ginn, G. and Coroneos, C. (2004) 'How to dig a dump: strategy and research design for investigation of Brisbane's nineteenth-century municipal dump', *Australasian Historical Archaeology*, 22: 15–26.

Hartmann, R., Lennon, J., Reynolds, D.P., Rice, A., Rosenbaum, A.T. and Stone, P.R. (2018) 'The history of dark tourism', *Journal of Tourism History*, 10(3): 269–295.

Heidelberg, B.A.W. (2015) 'Managing ghosts: exploring local government involvement in dark tourism', *Journal of Heritage Tourism*, 10(1): 74–90.

Herald (1928) 'Tourist value of Port Arthur ruins: case against demolition', *Herald*, 31 January: 10.

Heritage Council of Western Australia (2007) *Register of Heritage Places – Assessment Documentation*, Available from: http://inherit.stateherit age.wa.gov.au/Admin/api/file/169ef108-355c-0a71-f899-00d2eb96c691 [Accessed 2 December 2022].

Heritage Manager (2021) 'Go where you must go, and hope', *Museum of Fire*, 22 December, Available from: https://www.museumoffire.net/sin gle-post/go-where-you-must-go-and-hope [Accessed 14 September 2022].

Heritage Victoria (2014) *Assessment Report Endorsed by The Heritage Council of Victoria to Amend an Existing Registration,* VHR H1549, File no. 14/002851, HERMES no. 119.

Hill, E. (1934) 'The tombstone tree', *Sun*, 27 May: 43.

Hohenhaus, P. (2013) 'Commemorating and commodifying the Rwandan genocide: memorial sites in a politically difficult context', in L. White and E. Frew (eds) *Dark Tourism and Place Identity: Managing and Interpreting Dark Places*, London: Routledge, pp 142–155.

Hohenhaus, P. (2021) *Atlas of Dark Destinations: Explore the World of Dark Tourism*, London: Laurence King Publishing, Orion Publishing Co.

Hollinshead, K. (1992) '"White" gaze, "red" people – shadow visions: the disidentification of 'Indians' in cultural tourism', *Leisure Studies*, 11(1): 43–64.

Hudson, S. (2017) 'Old rural prisons: Victoria's best', *The Weekly Times*, 10 August.

Huppatz, B. (1999) 'Building offered as tourist attraction; the bank of death for sale', *Advertiser* [Adelaide, South Australia, Australia], 5 November: 013.

Hyde, B. (2014) 'New owners of infamous Snowtown bank say it is simply a house', *The Advertiser*, [online] 23 May, Available from: https://www.adel aidenow.com.au/news/south-australia/new-owners-of-infamous-snowt own-bank-say-it-is-simply-a-house/news-story/970ba9361b13ab244b82e 4d0a21900b5 [Accessed 26 May 2022].

Illustrated Sydney News (1891) 'Port Arthur, Tasmania: a trip from Hobart', *Illustrated Sydney News*, 23 May: 18.

Indigo Shire Council (2017a) *Beechworth Historic Courthouse*, Available from: https://www.explorebeechworth.com.au/listing/beechworth-histo ric-courthouse/ [Accessed 30 August 2022].

Indigo Shire Council (2017b) *Burke Museum*, Available from: https://www.explorebeechworth.com.au/burke-museum-and-historic-precinct/burke-museum/ [Accessed 30 August 2022].

Indigo Shire Council (2017c) *Tours & Public Programs*, Available from: https://www.explorebeechworth.com.au/burke-museum-and-historic-precinct/tours-public-programs [Accessed 30 August 2022].

Indigo Shire Council (2022a) *Beechworth Courthouse Kelly Trials Project*, Available from: https://www.indigoshire.vic.gov.au/Residents/Proje cts-works/Preserving-our-Heritage/Beechworth-Courthouse-Kelly-Tri als-Project [Accessed 30 August 2022].

Indigo Shire Council (2022b) *Ned Kelly Vault Exhibition on the Move*, Available from: https://www.indigoshire.vic.gov.au/About-Council/Latest-news/ Ned-Kelly-Vault-Exhibition-on-the-move [Accessed 31 August 2022].

Israfilova, F. and Khoo-Lattimore, C. (2019) 'Sad and violent but I enjoy it: children's engagement with dark tourism as an educational tool', *Tourism and Hospitality Research*, 19(4): 478–487.

Jackman, G. (2001) 'Get thee to church: hard work, godliness and tourism at Australia's first rural reformatory', *Australasian Historical Archaeology*, 19: 6–13.

Jobes, P.C., Barclay, E., Donnermeyer, J.F. and Graycar, A. (2001) 'Rural crime in Australia: contemporary concerns, recent research and future directions', *Australasian Journal of Regional Studies*, 7(1): 3–21.

Jones, J.K. (2016) *Historical Archaeology of Tourism at Port Arthur, Tasmania, 1885–1960*. PhD dissertation, Simon Fraser University.

Kate's Cottage (2022a) *The Kelly Homestead*, Available from: http://www.katescottageglenrowan.com.au/the-kelly-homestead/ [Accessed 30 August 2022].

Kate's Cottage (2022b) *Kate's Cottage*, Available from: http://www.katesc ottageglenrowan.com.au/kates-cottage/ [Accessed 30 August 2022].

Keane, D. and Martin, P. (2019) 'Life after death: dark tourism and the future of Snowtown', *ABC News*, [online] 20 May, Available from: https://www.abc.net.au/news/2019-05-20/can-snowtown-ever-shake-off-its-dark-past/11082778?nw=0 [Accessed 26 May 2022].

Kidron, C.A. (2013) 'Being there together: dark family tourism and the emotive experience of co-presence in the Holocaust past', *Annals of Tourism Research*, 41: 175–194.

Kincaid, J.R. (1997) 'Foreward', in Richard Tithecott, *Of Men and Monsters: Jeffrey Dahmer and the Construction of the Serial Killer'*, Madison, WI: The University of Wisconsin Press.

Kim, S. and Butler, G. (2015) 'Local community perspectives towards dark tourism development: the case of Snowtown, South Australia', *Journal of Tourism and Cultural Change* 13(1): 78–89.

Kimberley Land Waterfront Holiday Park (2022) 'Prison boab tree', *Kimberley Land Waterfront Holiday Park*, Available from: https://www.kimberleyland.com.au/things-to-do/prison-boab-tree [Accessed 1 March 2022].

Koehler, S.A., Moore, P. and Owen, D. (2009) *Jumped, Fell, or Pushed?: How Forensics Solved 50 'Perfect' Murders*, Chatswood, NSW: New Holland.

Lantern Ghost Tours (2021) *J Ward Lunatic Asylum Overnight Paranormal Investigation Victoria*, Available from: https://lanternghosttours.rezdy.com/99761/j-ward-lunatic-asylum-overnight-paranormal-investigation-victoria?currency=CAD [Accessed 8 December 2021].

Leader (1890) 'The ladies page: a trip to Port Arthur', *Leader*, 15 March: 5.

Lemelin, R.H., Whyte, K.P., Johansen, K., Desbiolles, F.H., Wilson, C. and Hemming, S. (2013) 'Conflicts, battlefields, indigenous peoples and tourism: addressing dissonant heritage in warfare tourism in Australia and North America in the twenty-first century', *International Journal of Culture, Tourism and Hospitality Research*, 7(3): 257–271.

Lennon, J. (2002) 'The Broad Arrow Café, Port Arthur, Tasmania: using social values methodology to resolve the commemoration issues', *Historic Environment*, 16(3): 38–46.

Lennon, J. (2017) 'Dark tourism sites: visualization, evidence and visitation', *Worldwide Hospitality and Tourism Themes*, 9(2): 216–227.

Lennon, J. and Foley, M. (2000) *Dark Tourism: The Attraction of Death and Disaster*, Boston, MA: Cengage Learning.

Lennon, J.J. (2010) 'Dark tourism and sites of crime', in D. Botterill and T. Jones (eds) *Tourism and crime*, Oxford: Goodfellow Publishers, pp 99–121.

Lennon, J.J. (2018) 'Dark tourism visualisation: Some reflections on the role of photography', in P.R. Stone, R. Hartmann, T. Seaton, R. Sharpley and L. White (eds) *The Palgrave Handbook of Dark Tourism Studies*, Palgrave: London, pp 585–602.

Lewis, R. (2017) 'Fringe review: Kapunda Ghost Crime Tour', *Glam Adelaide*, [online] 5 March, Available from: https://glamadelaide.com.au/fringe-review-kapunda-ghost-crime-tour/ [Accessed 22 November 2022].

LGIU (2023) *Poll: Help Shape Our Dark Tourism and Local Government Research*, [online] 23 October, Available from: https://lgiu.org/blog-article/poll-help-shape-our-dark-tourism-and-local-government-research/ [Accessed 12 February 2024].

Light, D. (2017) 'Progress in dark tourism and thanatourism research: an uneasy relationship with heritage tourism', *Tourism Management*, 61: 275–301.

Light Regional Council (2020) *Kapunda – Get Closer to History and Innovation*, Available from: https://www.lightcountry.com.au/kapunda [Accessed 22 November 2022].

Lippard, L.R. (1999) *On the Beaten Track: Tourism, Art and Place*, New York: New Press.

Lonely Planet (2022) *Ned Kelly Vault*, Available from: https://www.lonel yplanet.com/australia/victoria/beechworth/attractions/ned-kelly-vault/ a/poi-sig/1561648/362507 [Accessed 31 August 2022].

Lowe, P. (1998) *The Boab Tree*, Victoria: Lothian Books.

Lowenthal, D. (1975) 'Past time, present place: landscape and memory', *The Geographical Review*, 65(1): 1–36.

Lyons, E. (2020) 'Australia's Bushfire Towns Could See Wave of Dark Tourism', *10 Daily News*, [online] 9 January, Available from: https://10da ily.com.au/news/a200108smqxn/australias-bushfire-towns-could-see-wave-of-dark-tourism-in-future-20200109 [Accessed 4 August 2020].

MacCannell, D. (1989) *The Tourist: A New Theory of the Leisure Class*. New York: Schocken Books.

MacDonald, L. (2023) 'Extreme weather linked to climate change the "number one threat" to natural World Heritage sites, research finds', *ABC News*, [online] 28 July, Available from: https://www.abc.net.au/news/ 2023-07-28/climate-biggest-risk-to-world-heritage-areas-research-finds/ 102656800 [Accessed 28 July 2023].

Maguire, S. (2003) 'Made infamous by the bodies-in-the-barrels murders, residents try to resurrect fortunes of ailing town; Snowtown's fight to shed terrible stigma', *Advertiser* [Adelaide, South Australia, Australia], 23 August: 016.

Manderson, L. (2008) 'Acts of remembrance: the power of memorial and the healing of indigenous Australia', *Adler Museum Bulletin*, 34(2): 5–19.

Martini, A. and Buda, D.M. (2020) 'Dark tourism and affect: framing places of death and disaster', *Current Issues in Tourism*, 23(6): 679–692.

Marton, Z., Ernszt, I., and Birkner, Z. (2020) 'Holidays to the hells of earth – taking risk as a dark tourist?' *Deturope*, 12(1): 136–153.

Mason, R., Myers, D. and de la Torre, M. (2003) *Port Arthur Historic Site: Port Arthur Historic Site Management Authority*, Los Angeles: The Getty Conservation Institute.

Maxwell-Stewart, H. (2013) '"The Lottery of Life": convict tourism at Port Arthur Historic Site, Australia', *Prison Service Journal*, 10: 24–28.

Maxwell-Stewart, H. and Hood, S. (2010) *Pack of Thieves? 52 Port Arthur Lives*. Tasmania: Port Arthur Historic Site Management Authority.

Maxwell-Stewart, H. and Nicholson, L. (2017) 'Penal transportation, family history, and convict tourism', in J.Z. Wilson, S. Hodgkinson, J. Piché and K. Walby (eds) *The Palgrave Handbook of Prison Tourism*, London: Palgrave, pp 713–734.

Maxwell-Stewart, H. and Oxley, D. (2020) 'Convicts and the colonisation of Australia, 1788–1868', *Digital Panopticon*, Available from: https:// www.digitalpanopticon.org/Convicts_and_the_Colonisation_of_Austra lia,_1788-1868 [Accessed 19 November 2021].

McCrossin's Mill Museum (2010) 'Thunderbolt life and legend', *McCrossin's Mill Museum*, Available from: https://uhs.org.au/2010/05/03/thunderbolt/ [Accessed 7 September 2022].

McDonald, W. and Davies, K. (2015) 'Creating history: literary journalism and Ned Kelly's last stand', *Australian Journalism Review,* 37(2): 33–49.

McGowan, M. (2019) 'Ivan Milat's chilling serial backpacker murders still haunt Australia', *The Guardian*, [online] 27 October, Available from: https://www.theguardian.com/australia-news/2019/oct/27/ivan-milat-chilling-serial-murders-haunt-australia-after-death [Accessed 3 May 2022].

McKenna, M. (2002) *Looking for Blackfellas' Point: An Australian History of Place*. Sydney: UNSW Press.

McKercher, B. (2001) 'Attitudes to a non-viable community-owned heritage tourist attraction', *Journal of Sustainable Tourism*, 9(1): 29–43.

McKercher, B. and du Cros, H. (2002) *Cultural Tourism: The Partnership between Tourism and Cultural Heritage Management*, Abingdon: Routledge.

Mercury (1880) 'Excursion to Port Arthur', *Mercury*, 30 March: 3.

Mercury (1880) 'Excursion to Port Arthur', *Mercury*, 10 November: 2.

Mercury (1881) 'Excursions to Port Arthur', *Mercury*, 28 January: 2.

Mercury (1900) 'Down to Port Arthur', *Mercury*, 27 December: 3.

Mercury (1939) 'History in stone: Plans for Port Arthur, guides' tales', *Mercury*, 8 December: 2.

Miller, R. (2019) 'Should we really be ok with a Jeffrey Dahmer tour in Milwaukee?', *Milwaukee*, Available from: https://www.milwaukeemag.com/really-ok-jeffrey-dahmer-tour-milwaukee/ [Accessed 31 May 2022].

Mining Editor (2014) 'Blue murder at Wittenoom', *Australian Mine Safety Journal*, [online] 6 May, Available from: https://www.amsj.com.au/blue-murder-wittenoom/ [Accessed 28 September 2022].

Monument Australia (2022) *Frontier*, Available from: https://monumentaustralia.org.au/themes/conflict/frontier [Accessed 30 November 2022].

Museums and Galleries NSW (2021) 'Berrima Courthouse museum', *Museums and Galleries of NSW*, Available from: https://mgnsw.org.au/organisations/berrima-courthouse-museum/ [Accessed 13 December 2021].

Museums of History NSW (2023) 'Mugshot height chart', *Museums of History NSW*, Available from: https://shop.mhnsw.au/products/mugshot heightchart?_pos=1&_sid=877328344&_ss=r [Accessed 15 May 2023].

MyPokeCard (2019) *Pokémon Captain Thunderbolt*, Available from: https://www.mypokecard.com/en/Gallery/Pokemon-Captain-Thunderbolt [Accessed 9 September 2022].

National Museum of Australia (2022) *Exhibitions: Great Southern Land*, Available from: https://www.nma.gov.au/exhibitions/great-southern-land [Accessed 22 November 2022].

National Parks (2021a) *Walk on the Dark Side: Sunset Tour*, Available from: https://www.nationalparks.nsw.gov.au/things-to-do/guided-tours/walk-on-the-dark-side-sunset-tour [Accessed 7 December 2021].

National Parks (2021b) *Trial Bay Gaol: Life Behind Bars Kids Tour*, Available from: https://www.nationalparks.nsw.gov.au/things-to-do/guided-tours/trial-bay-gaol-life-behind-bars-kids-tour [Accessed 7 December 2021].

National Parks (2021c) *Trial Bay Gaol Twilight Tour*, Available from: https://www.nationalparks.nsw.gov.au/things-to-do/guided-tours/trial-bay-gaol-twilight-tour [Accessed 7 December 2021].

National Parks (2021d) *Trial Bay Gaol*, Available from: https://www.nationalparks.nsw.gov.au/things-to-do/historic-buildings-places/trial-bay-gaol?p=1&pdfprint=true [Accessed 7 December 2021].

Ned Kelly Touring Route (2022) *Eldorado and the Woolshed Valley*, Available from: https://nedkellytouringroute.com.au/destinations/eldorado-woolshed-valley/ [Accessed 3 August 2022].

Nettlebeck, A. (2011) 'The Australian frontier in the museum', *Journal of Social History*, 41(4): 1115–1128.

Newton, A. (2021) 'People urged not to travel to Wittenoom … ever', *Haveagonews*, [online] 7 September, Available from: https://www.haveagonews.com.au/news/people-urged-not-to-travel-to-wittenoom-ever/ [Accessed 28 September 2022].

Newton, M. (2006) *The Encyclopaedia of Serial Killers* (2nd edn), New York: Checkmark Books, pp 242–244.

Nichols, D. and Freeman, C.G. (2021) 'The importance of Australia's big things', *Pursuit*, 4 January, Available from: https://pursuit.unimelb.edu.au/articles/the-importance-of-australia-s-big-things [Accessed 27 September 2023].

Noonan, A. (2012) 'eBay pulls "death" bank from sale site', *Advertiser* [Adelaide, South Australia, Australia], 22 February: 12.

Noonan, A. (2012) 'There's money in the bank for Snowtown', *Advertiser* [Adelaide, South Australia, Australia], 3 March: 41.

O'Connell, R. (2020) '10 Reasons this Australian state is the ultimate offbeat destination', *FodorsTravel*, [online] 21 December, Available from: https://www.fodors.com/world/australia-and-the-pacific/australia/western-australia/experiences/news/10-reasons-this-australian-state-is-the-ultimate-offbeat-destination [Accessed 28 September 2022].

Officer Travels (2017) *Trial Bay Goal*, Available from: https://officertravels.com/trial-bay-gaol/ [Accessed 7 December 2021].

Old Beechworth Gaol (nd) 'Home', *Old Beechworth Gaol*, Available from: https://oldbeechworthgaol.com.au [Accessed 10 December 2021].

Old Melbourne Gaol (2022) *Hangman's Night Tour*, Available from: https://www.oldmelbournegaol.com.au/event/hangmans-night-tour/ [Accessed 31 August 2022].

O'Neill, S. (2002) 'Soham pleads with trippers to stay away', *Daily Telegraph*, August 26.

Only Melbourne (nd) 'J Ward lunatic asylum', *Only Melbourne*, Available from: https://www.onlymelbourne.com.au/j-ward-lunatic-asylum [Accessed 7 December 2021].

Owen, C. (2003) '"The police appear to be a useless lot up there": law and order in the East Kimberley 1884–1905', *Aboriginal History*, 27: 105–130.

Panayotov, J. (2023) 'Queensland Police Museum', *Must Do Brisbane*, Available from: https://www.mustdobrisbane.com/visitor-info-arts-culture-muse ums/queensland-police-museum-brisbane [Accessed 15 May 2023].

Parke, E. (2022) 'Judges go bush to learn about Indigenous culture, with aim to deliver fairer justice in courts', *ABC News*, [online] 14 September, Available from: https://www.abc.net.au/news/2022-09-14/judges-go-bush-in-wa-to-learn-about-indigenous-culture/101414554 [Accessed 30 November 2022].

Parliament of Tasmania (2020) *Legislative Council Government Business Scrutiny Committee B, Port Arthur Historic Site Management Authority*, 15 December, Available from: https://www.parliament.tas.gov.au/ctee/Council/Transcri pts/GBE%202020/LC%20GBE%20Tuesday%2015%20December%202 020%20-%20PAHSMA.pdf [Accessed 20 October 2021].

Pascoe, R. (2020) 'Black Summer and Beyond photo exhibition highlights the strength, bravery and resilience of the Macleay', *The Macleay Argus*, [online] 9 November, Available from: https://www.macleayargus.com.au/story/7002 545/black-summer-and-beyond-photo-exhibition-highlights-the-strength-bravery-and-resilience-of-the-macleay/ [Accessed 22 November 2022].

Patty, A. (2015) 'Belanglo State Forest "extreme terror tour" angers families of murder victims', *The Sydney Morning Herald*, [online] 13 July, Available from: https://www.smh.com.au/national/nsw/belanglo-state-forest-extr eme-terror-tour-angers-families-of-murder-victims-20150712-giadmu. html [Accessed 3 May 2022].

Pearce, P.L., Morrison, A.M. and Moscardo, G.M. (2003) 'Individuals as tourist icons: a developmental and marketing analysis', *Journal of Hospitality & Leisure Marketing*, 10(1–2): 63–85.

Peters, A. and Higgins-Desbioelles, F. (2012) 'De-marginalising tourism research: indigenous Australians as tourists', *Journal of Tourism and Hospitality Management*, 19: 1–9.

Phelps, J. (2023) 'Mystery of Ned Kelly skull revealed at last', *Australian*, 24 September.

Phillips, K. (1999) 'Three months after having international fame or infamy thrust upon it by the missing people killings, the citizens of the rural community of Snowtown are gradually getting their disrupted lives back in order Snowtown throws off ...', *Advertiser* [Adelaide, South Australia, Australia], 9 August: 012.

Piché, J. and Walby, K. (2010) 'Problematizing carceral tours', *The British Journal of Criminology*, 50(3): 570–581.

Piché, J. and Walby, K. (2018) 'Dark tourism, penal landscapes, and criminological inquiry', in N. Rafter and M. Brown (eds) *The Oxford Encyclopaedia of Crime, Media, and Popular Culture* (volume 1: A–E), Oxford: Oxford University Press, pp 563–575.

Piper, A.K.S. (2016) 'The future of the past – a Cautionary lesson: heritage and financial mismanagement at the Port Arthur Historic Site, 1987–1996' *Australian Folklore*, 31: 237–260.

Podoshen, J.S., Venkatesh, V., Wallin, J., Andrzejewski, S.A. and Jin, Z. (2015) 'Dystopian dark tourism: an exploratory examination', *Tourism Management*, 51: 316–328.

Port Arthur Historic Site (PAHS) (2019) *History Timeline*, Available from: https://portarthur.org.au/history/history-timeline/ [Accessed: 10 May 2019].

Port Arthur Historic Site Management Authority (PAHSMA) (2009) *Port Arthur Historic Sites Statutory Management Plan 2008*, Available from: https://portarthur.org.au/wp-content/uploads/2017/11/SMP_APRIL_2009.pdf [Accessed 10 May 2019].

Port Arthur Historic Site Management Authority (PAHSMA) (2018) *Annual Report 2017–18*, Port Arthur Historic Site Management Authority (PAHSMA).

Port Arthur Historic Site Management Authority (PAHSMA) (2019) *Heritage Values*, Available from: https://portarthur.org.au/heritage-management/heritage-values [Accessed 19 November 2021].

Port Arthur Historic Site Management Authority (PAHSMA) (nd) *Port Arthur Memorial Garden Brochure*, Port Arthur Historic Site Management Authority (PAHSMA), Available from: https://portarthur.org.au/wp-content/uploads/2018/01/Port-Arthur-Memorial-Garden-Brochure1.pdf [Accessed 19 November 2021].

Pridmore, W.B. (2009) *Port Arthur ... Convicts and Commandants*, Melbourne: Graphic Impressions.

Priestland, A. (2018) 'Ned Kelly's skull is still missing, and the mystery continues to grow 138 years after his death', *ABC News*, [online] 11 November, Available from: https://www.abc.net.au/news/2018-11-11/ned-kelly-skull-location-remains-a-mystery/10476666 [Accessed 31 August 2022].

Queenslander (1931) 'The baobab, or Western Australian bottle tree', *Queenslander*, 26 February: 29.

Raymen, T. and Smith, O. (2019) 'Introduction: why leisure?', in T. Raymen and O. Smith (eds) *Deviant Leisure: Criminological Perspectives on Leisure and Harm*, Palgrave MacMillan, pp 1–13.

Reid, A., Berry, G., de Klerk, N., Hansen, J., Heyworth, J., Ambrosini, G., Fritschi, L., Olsen, N., Merler, E. and Musk A.W. (2007) 'Age and sex differences in malignant mesothelioma after residential exposure to blue asbestos (Crocidolite)', *Chest*, 131: 376–382.

Ribber, S. (1940) 'A "boob" in a boab tree: a queer lock-up', *The Sydney Morning Herald*, 31 August: 9.

Robb, E.M. (2009) 'Violence and recreation: vacationing in the realm of dark tourism', *Anthropology and Humanism*, 34(1): 51–60.

Rofe, M.W. (2013) 'Considering the limits of rural place making opportunities: rural dystopias and dark tourism', *Landscape Research*, 38(2): 262–272.

Rojek C. (1993) *Ways of Escape*, Basingstoke: Macmillan.

Rolph. R.S. (1935) 'Built upon a memory: tourist resort at old Port Arthur', *Examiner*, 20 July: 4.

Rosser, E. (2013) 'A place for monsters: *Wolf Creek* and the Australian outback', *Monsters and the Monstrous* 3(2): 73–82.

Rural City of Wangaratta (2022) *Glenrowan*, Available from: https://www.visitwangaratta.com.au/Discover-the-region/Glenrowan [Accessed 31 August 2022].

Ryan So Fly (2021) 'Travel review: KAPUNDA – Australia's most haunted town', *YouTube*, 16 December, Available from: https://www.youtube.com/watch?v=YvtnZsHSUxg [Accessed 22 November 2022].

Sampson, H. (2019) 'By the way dark tourism, explained: why visitors flock to sites of tragedy', *Washington Post*, [online] 13 November, Available from: https://www.washingtonpost.com/graphics/2019/travel/dark-tourism-explainer/ [Accessed 19 November 2021].

Scanlon, M. (2019) 'History: Trial Bay Gaol a model of remote control', *Newcastle Herald*, [online] 12 January, Available from: https://www.newcastleherald.com.au/story/5843790/resorting-to-remote-control/ [Accessed 7 December 2021].

Schechter, H. and Everitt, D. (2006) *The A to Z Encyclopedia of Serial Killers*. New York: Gallery Books.

Scott, J. and Biron, D. (2010) 'Wolf Creek, rurality and the Australian gothic', *Continuum*, 24(2): 307–322.

Scott, J. and Hogg, R. (2015) 'Strange and stranger ruralities: social constructions of rural crime in Australia', *Journal of Rural Studies*, 39: 171–179.

Scott, K. (2019) 'A warning to tourists still flocking to Australia's most deadly town', *Nine*, [online] 10 July, Available from: https://travel.nine.com.au/latest/australias-most-contaminated-town-wittenoon-abandoned/b7752071-b209-452a-bdfb-442a73b66c25 [Accessed 28 September 2022].

Scott, R. and MacFarlane, I. (2014) 'Ned Kelly – stock thief, bank robber, murderer – psychopath', *Psychiatry, Psychology and Law*, 21(5): 716–746.

Scully, M.A. (2021) 'Join the team at Old Beechworth Gaol', *ACRE*, Available from: https://acre.org.au/join-the-team-at-old-beechworth-gaol/ [Accessed 10 December 2021].

Seaton A. V. (1996) 'Guided by the dark: from thanatopsis to thanatourism', *Journal of Heritage Studies*, 2(4): 234–244.

Seaton, A.V. and Lennon, J.J. (2004) 'Thanatourism in the early 21st century: moral panic, ulterior motives and alterior desires', in T.V. Singh (ed) *New Horizons in Tourism: Strange Experiences and Stranger Practices*, Wallingford: CAB International, pp 63–82.

Serventy, V. (1966) 'Dance of the boab', *The Australian Women's Weekly*, 2 February: 26.

Serventy, V. (1967) *Nature Walkabout*, Artarmon: AH and AW Reed.

Sharpley, R. (2005) 'Travels to the edge of darkness: towards a typology of "dark tourism"', in M. Aicken, S.J. Page and C. Ryan (eds) *Taking Tourism to the Limits: Issues, Concepts and Managerial Perspectives*. Oxford: Elsevier, pp 215–226.

Sharpley, R. (2009) 'Dark tourism and political ideology: towards a governance model', in R. Sharpley and P.R. Stone (eds) *The Darker Side of Travel: The Theory and Practice of Dark Tourism*. Bristol: Channel View, pp 145–163.

Sharpley, R. and Stone, P.R. (2009) 'Life, death and dark tourism: future research directions and concluding comments', in R. Sharpley and P.R. Stone (eds), *The Darker Side of Travel: The Theory and Practice of Dark Tourism*. Bristol: Channel View, pp 247–251.

Sharpley, R. and Wright, D. (2018) 'Disasters and disaster tourism: the role of the media', in P.R. Stone, R. Hartmann, T. Seaton, R. Sharpley and L. White (eds) *The Palgrave Handbook of Dark Tourism Studies*, London, Palgrave Macmillan, pp 335–354.

Shehata, W., Langston, C. and Sarvimaki, M. (2018) 'From uncomfortable to comfortable: the adaptive reuse of Australian gaols', *International Heritage and Cultural Conservation Conference*, 3–5 December 2018, Sarawak.

Sherri (2017) 'Boab prison tree, Wyndham', *Kimberley Croc Motel*, Available from: https://www.kimberleycrocmotel.com.au/boab-prison-tree-wynd ham/ [Accessed 7 December 2022].

Silvester, J. (2018) 'Ned Kelly: setting the story straight at Stringybark Creek', *The Age*, [online] 6 December, Available from: https://www.theage.com. au/national/victoria/ned-kelly-setting-the-story-straight-at-stringybark-creek-20181205-p50kck.html [Accessed 31 August 2022].

Smith, B. (2009) *The Startling Legacy of the Convict Era*. Allen & Unwin.

Smith, R. (2015) 'Dark tourism proves Australians are fascinated with mass murder, violence and torture', *News*, [online] 13 August, Available from: https://www.news.com.au/travel/travel-ideas/adventure/dark-tourism-proves-australians-are-fascinated-with-mass-murder-violence-and-torture/news-story/eebc9a6b8855c2c9caedb6a883663e21 [Accessed 4 August 2020].

Smith, W. (2002) 'Descent into depravity', *The Courier-Mail* (Brisbane), 10 May: 34.

Smyrk, K. (2022) 'Aboriginal artefacts to return home to South Australia's Kaurna people after 150 years in Beechworth museum', *ABC News,* [online] 3 July, Available from: https://www.abc.net.au/news/2022-07-03/aboriginal-artefacts-return-kaurna-people-sa/101203712 [Accessed 30 November 2022].

Somerville, E. (2017) 'Site of Kelly gang shootout with police at Stringybark Creek being prepared for 140th anniversary', *ABC News*, [online] 27 March, Available from: https://www.abc.net.au/news/2017-03-27/site-of-kelly-gang-shootout-with-police-at-stringybark-creek/8389678 [Accessed 31 August 2022].

South Western Times (1923) 'Life at Wyndham', *South Western Times*, 5 June: 2.

Spero, P. (1939) 'Thunderbolt's' grave', *Western Mail* (Perth), 10 August: 10.

Staples, M. (1995) 'Heritage, tourism and local communities', *Rural Society* 5(1): 35–40.

State Government of Victoria (2021) 'Kelly case files: exploring the Kelly legend', *Victorian Police Museum*, Available from: https://www.policemuseum.vic.gov.au/kelly-case-files [Accessed 31 August 2022].

State Library of New South Wales (2022) *Bushrangers of New South Wales: Captain Thunderbolt*, Available from: https://www.sl.nsw.gov.au/stories/bushrangers-new-south-wales/captain-thunderbolt [Accessed 7 September 2022].

State Library of South Australia (2009) 'Photographs from M.E. McCombe's nursing days [PRG 900/6/3]', Available from: https://collections.slsa.sa.gov.au/resource/PRG+900/6/3 [Accessed 29 November 2022].

Sticky Tickets (2020) 'Berrima Courthouse HALLOWEEN Ghost Tour', *Sticky Tickets*, Available from: https://www.stickytickets.com.au/xh0gl/berrima_courthouse_halloween_ghost_tour.aspx, [Accessed 12 May 2023].

Stone, P. (2006) 'A dark tourism spectrum: towards a typology of death and macabre related tourist sites, attractions and exhibitions', *Tourism,* 54(2): 145–160.

Stone, P.R. (2009a) 'Making absent death present: consuming dark tourism in contemporary society', in R. Sharpley and P.R. Stone (eds) *The Darker Side of Travel: The Theory and Practice of Dark Tourism*, Bristol: Channel View, pp 23–38.

Stone, P.R. (2009b) 'Dark tourism: morality and new moral spaces', in R. Sharpley and P.R. Stone (eds) *The Darker Side of Travel: The Theory and Practice of Dark Tourism*, Bristol: Channel View, pp 56–72.

Stone, P.R. (2013) 'Dark tourism, heterotopias and post-apocalyptic places: the case of Chernobyl', in L. White and E. Frew (eds) *Dark Tourism and Place Identity*, Abingdon: Routledge, pp 79–93.

Stone, P.R. (2017) 'Ethics of dark tourism: towards a model of morality in secular society', *Current Issues in Dark Tourism Research*, E0002-2017-PS, pp 1–17.

Strahan, C. (2018) 'Policemen killed in Kelly Gang shootout honoured at Stringybark Creek memorial for 140th anniversary', *ABC News*, [online] 10 December, Available from: https://www.abc.net.au/news/2018-12-10/policemen-killed-in-kelly-gang-shootout-honoured-at-stringybark/10600 870 [Accessed 31 August 2022].

Strange, C. (2000a) 'From "Place of Misery" to "Lottery of Life": interpreting Port Arthur's past', *Open Museum Journal*, 2 (*Unsavoury Histories*).

Strange, C. (2000b) 'The Port Arthur massacre: tragedy and public memory in Australia', *Studies in Law, Politics and Society*, 20: 159–182.

Strange C. and Kempa, M. (2003) 'Shades of dark tourism: Alcatraz and Robben Island', *Annals of Tourism Research*, 30: 386–405.

Sunday Times (1919) 'A baobab tree at the King River Pool, Kimberley', *Sunday Times*, 2 February: 9.

Sutton, C. (2020) 'Who was the mother of the most depraved serial killer of all time?', *News.com.au*, [online] 25 October, Available from: https://www.news.com.au/world/north-america/who-was-the-mother-of-the-most-depraved-serial-killer-of-all-time/news-story/5320e1f04b061279c 46af1ba67601edd [Accessed 10 May 2021].

Tasmania Police Museum (nd) 'Current project', *Tasmania Police Museum*, Available from: https://www.tasmaniapolicemuseum.com.au/current-proj ect, [Accessed 15 May 2023].

Tenterfield Shire Council (2022) 'Captain Thunderbolt', *Visit Tenterfield*, Available from: https://www.visittenterfield.com.au/things-to-do/history-heritage/famous-faces/captain-thunderbolt [Accessed 6 September 2022].

The Sun (1919) 'Trial Bay Monument: German landmark destroyed', *The Sun*, 11 July: 1.

The Urana Independent and Clear Hills Standard (1919) 'German monument', *The Urana Independent and Clear Hills Standard*, 11 July: 1.

The West Australian (2021) 'Wittenoom visitors dicing with death', *The West Australian*, [online] 3 September, Available from: https://thewest.com.au/travel/wa/wittenoom-visitors-dicing-with-death-ng-b8819920 29z [Accessed 28 September 2022].

Thompson, R. (2015) 'NSW small business cancels Ivan Milat-themes ghost tours after backlash', *Smart Company*, [online] 15 July, Available from: https://www.smartcompany.com.au/finance/nsw-small-busin ess-cancels-ivan-milat-themed-ghost-tours-after-backlash/ [Accessed 3 May 2022].

Tinson, J.S., Saren, M.A.J. and Roth, B.E. (2015) 'Exploring the role of dark tourism in the creation of national identity of young Americans', *Journal of Marketing Management*, 31(7–8): 856–880.

Towie, N. (2022) ' "I'm so angry, I'm wild": the never-ending wait to clean up asbestos town Wittenoom', *The Guardian*, [online] 30 May, Available from: https://www.theguardian.com/environment/2022/may/30/im-so-angry-im-wild-the-never-ending-wait-to-clean-up-asbestos-town-wittenoom [Accessed 28 September 2022].

Tranter, B. and Donoghue, J. (2010) 'Ned Kelly: armoured icon', *Journal of Sociology*, 46(2): 187–205.

Tranter, B. and Donoghue, J. (2008) 'Bushrangers: Ned Kelly and Australian identity', *Journal of Sociology*, 44(4): 373–390.

Tribune (1877) 'Abolition of Port Arthur', *Tribune*, 18 April: 2.

Tribune (1877) 'Trip to Port Arthur', *Tribune*, 27 December: 3.

Tripadvisor (2021a) 'Old Beechworth Gaol', *Tripadvisor*, Available from: https://www.tripadvisor.com.au/Attraction_Review-g552137-d2254316-Reviews-Old_Beechworth_Gaol-Beechworth_Victoria.html [Accessed 10 December 2021].

Tripadvisor (2021b) 'Old Dubbo Gaol', *Tripadvisor*, Available from: https://www.tripadvisor.com.au/Attraction_Review-g255320-d2097439-Reviews-Old_Dubbo_Gaol-Dubbo_New_South_Wales.html [Accessed 14 December 2021].

Tripadvisor (2021c) 'Berrima Courthouse', *Tripadvisor*, Available from: https://www.tripadvisor.com.au/Attraction_Review-g528923-d574850-Reviews-or20-Berrima_Courthouse-Berrima_Southern_Highlands_New_South_Wales.html [Accessed 13 December 2021].

Tripadvisor (2022) 'Ghost crime tours', *Tripadvisor*, Available from: https://www.tripadvisor.com.au/Attraction_Review-g255093-d7058013-Reviews-Ghost_Crime_Tours-Adelaide_Greater_Adelaide_South_Australia.html [Accessed 22 November 2022].

Tripadvisor (2023) 'Timber Creek Police Station and Museum', *Tripadvisor*, Available from: https://www.tripadvisor.com.au/Attraction_Review-g494975-d3914003-Reviews-Timber_Creek_Police_Station_and_Museum-Timber_Creek_Top_End_Northern_Territory.html [Accessed 15 May 2023].

Tripadvisor (2024) 'Explore Milwaukee', *Tripadvisor*, Available from: https://www.tripadvisor.com/Tourism-g60097-Milwaukee_Wisconsin-Vacations.html [Accessed 13 February 2024].

Tucker, H. and Akama, J. (2009) 'Tourism as postcolonialism', in T. Jamal and M. Robinson (eds) *SAGE Handbook of Tourism Studies*, London: SAGE, pp 505–521 (online pages 1–15).

Tucker, H., Shelton, E.J. and Bae, H. (2017) 'Post-disaster tourism: towards a tourism of transition', *Tourist Studies*, 17(3): 306–327.

Tumarkin, M. (2005) *Traumascapes: The Power and Fate of Places Transformed by Tragedy*. Melbourne: Melbourne University Press.

Tunbridge, J.E. and Ashworth, G.J. (1996) *Dissonant Heritage: The Management of the Past as a Resource in Conflict*. New York: Wiley.

Twain, M. (2003) *The Innocents Abroad*, New York: Random House.

Twyford-Moore, S. (2017) 'The trees [online]', *Kill Your Darlings*, 29: 144–151.

Uralla Times and District Advocate (1919) 'Thunderbolt's grave', *Uralla Times and District Advocate*, 3 May: 2.

Uralla Visitor Information Centre (nd) *Captain Thunderbolt*, Available from: https://www.uralla.com/files/assets/urallacom/vic-info-pdfs/captain-thunderbolt.pdf [Accessed 7 September 2022].

Virgin Australia (2016) 'A haunted tour of South Australia, 8 macabre experiences', *South Australia Traveller*, Available from: https://www.virginaustralia.com/inspiration/au/travel-tips/haunted-tour-south-australia-8-macabre-experiences [Accessed 22 November 2022].

Visit Melbourne (2022) *Ned Kelly: A Brief but Remarkable Life*, Available from: https://www.visitmelbourne.com/regions/high-country/see-and-do/art-and-culture/history-and-heritage/bushranger-history/ned-kelly [Accessed 10 August 2022].

Wakelin, J. (1999) 'No laughing matter: Snowtown merchandise is sick, says victim's mother', *Advertiser* [Adelaide, South Australia, Australia], 13 October: 006.

Wakelin, J. and Oakley, V. (1999) 'Selling of slaughter: macabre souvenirs on the market in Snowtown', *Advertiser* [Adelaide, South Australia, Australia], 12 October 1999: 1 and 4.

Walby, K. and Piché, J. (2011) 'The polysemy of punishment memorialization: dark tourism and Ontario's penal history museums', *Punishment & Society*, 13(4): 451–472.

Walmsley, D.J. (2003) 'Rural tourism: a case of lifestyle-led opportunities', *Australian Geographer*, 34(1): 61–72.

Ward, R. (2003) *The Australian Legend*, Victoria: Oxford University Press.

Weidenhofer, M. (1990) *Port Arthur: A Place of Misery*, Tasmania: Port Arthur Historic Site Management Authority.

Weir, S. (2000) 'Snowtown locals get together to brush off a tarnished image: a town fights back', *Advertiser* [Adelaide, South Australia, Australia], 5 January: 030.

Welch, M. (2012) 'Penal tourism and the "dream of order": exhibiting early penology in Argentina and Australia', *Punishment & Society*, 14(5): 584–615.

Welch, M. (2013) 'Penal tourism and a tale of four cities: reflecting on the museum effect in London, Sydney, Melbourne, and Buenos Aires', *Criminology & Criminal Justice*, 13(5): 479–505.

Western Australia (nd) 'Attraction: boab prison tree', *Welcome to Western Australia*, Available from: https://www.westernaustralia.com/en/Attraction/Boab_Prison_Tree/56b26774aeeeaaf773cfa415 [Accessed 1 March 2022].

Western Australia, Legislative Council (2021) *Wittenoom Closure Bill 2021*, 27 October 2021: 4877c–4878a.

Western Herald (1965) 'Historical buildings', *Western Herald*, 16 July: 2.

Wheaton, C. and Reardon, A. (2020) 'National Museum's Black Summer exhibition to preserve memories of Australia's bushfire crisis', *ABC News*, [online] 2 April, Available from: https://www.abc.net.au/news/2020-04-02/black-summer-debris-set-for-national-museum-exhibition/12114144 [Accessed 13 September 2022].

White, R. (2016) 'From trauma to tourism and back again: Port Arthur's history of "dark tourism"', *The Conversation*, [online] 27 April, Available from: https://theconversation.com/from-trauma-to-tourism-and-back-again-port-arthurs-history-of-dark-tourism-56993 [Accessed 19 November 2021].

Whitford, M. and Ruhanen, L. (2016) 'Indigenous tourism research, past and present: where to from here?', *Journal of Sustainable Tourism*, 24(8-9): 1080–1099.

Whitley, D. (2020) 'Is the Ned Kelly Tree at Stringybark Creek real?', *Australia Travel Questions*, [online] 4 November, Available from: https://australiatravelquestions.com/history/ned-kelly-tree-stringybark-creek/ [Accessed 31 August 2022].

Wickens, G. and Lowe, P. (2008) *The Baobabs: Pachycauls of Africa, Madagascar and Australia*, New York: Springer.

Williams P. (2007) *Memorial Museums: The Global Rush to Commemorate Atrocities*. Oxford: Berg.

Willis, B. (2018) 'Coming back to Snowtown', *Advertiser* [Adelaide, South Australia, Australia], 24 June: 28.

Wilson, C. and Odell, R. (1987) *Jack the Ripper: Summing Up and Verdict*, London; New York: Bantam.

Wilson, J.Z. (2004) 'Dark tourism and the celebrity prisoner: front and back regions in representations of an Australian historical prison', *Journal of Australian Studies*, 28(82): 1–13.

Wilson, J.Z. (2008a) *Prison: Cultural Memory and Dark Tourism*, Lausanne: Peter Lang.

Wilson, J.Z. (2008b) 'Transgressive decor: narrative glimpses in Australian prisons, 1970s–1990s', *Crime, Media, Culture*, 4(3): 331–348.

Wilson, J.Z. (2011a) 'Australian prison tourism: a question of narrative integrity', *History Compass*, 9: 562–571.

Wilson, J.Z. (2011b) 'Dark tourism and national identity in the Australian history curriculum', in E. Frew and L. White (eds) *Tourism and National Identity: An International Perspective*, Abingdon: Routledge, pp 202–214.

Wilson, J.Z., Hodgkinson, S., Piché, J. and Walby, K. (2017) 'Introduction: prison tourism in context', in J.Z. Wilson, S. Hodgkinson, J. Piché and K. Walby (eds) *The Palgrave Handbook of Prison Tourism*, London: Palgrave, pp 1–10.

Wise, J. and McLean, L. (2021) 'Making light of convicts: branding "bubbly" with offender images', *M/C Journal*, 24(1), Available from: https://journal.media-culture.org.au/index.php/mcjournal/article/view/2737 [Accessed 6 May 2024].

Wise, J. and Roberts, D.A. (2016) 'Developments of crime and criminal justice system in Australia', in A. Harkness, B. Harris and D. Baker *Locating Crime in Context and Place: Perspectives on Regional, Rural and Remote Australia*, Sydney: Federation Press, pp 35–48.

Witcomb, A. (2013) 'Using immersive and interactive approaches to interpreting traumatic experiences for tourists: potentials and limitations', in R. Staiff, R. Bushell and S. Watson (eds) *Heritage and Tourism: Place, Encounter, Engagement*, Abingdon: Routledge, pp 152–170.

Woolf, M. (2023) 'The rise of dark tourism [2022 study]', *Passport Photo Online*. [online] 7 July, Available from: https://passport-photo.online/blog/rise-of-dark-tourism/ [Accessed 12 February 2024].

World (1919) 'Peace at Port Arthur: a tourist's impression', *World*, 4 January: 3.

Wright, D. and Sharpley, R. (2018) 'The photograph: tourist responses to a visual interpretation of a disaster', *Tourism Recreation Research*, 43(2): 161–174.

Wright, T. (1988) 'Thunderbolt – vicious criminal or singing cowboy?' *The Canberra Times*, 28 February: 3.

Young, D. (1996) *Making Crime Pay: The Evolution of Convict Tourism in Tasmania*. Sandy Bay: Tasmanian Historical Research Association.

Zaunmayr, T. (2019) 'Fears Insta fame driving tourists to WA's deadly cancer town', *The West Australian*, [online] 16 March, Available from: https://thewest.com.au/news/regional/fears-insta-fame-driving-tourists-to-was-deadly-cancer-town-ng-b881129037z [Accessed 28 September 2022].

Zhang, Y. (2022) 'Experiencing human identity at dark tourism sites of natural disasters', *Tourism Management*, 89(104451): 1–11.

# Index